Andrew Steinmetz

Sunshine and Showers

Their Tnfluences throughout Creation

Andrew Steinmetz

Sunshine and Showers
Their Tnfluences throughout Creation

ISBN/EAN: 9783337216016

Printed in Europe, USA, Canada, Australia, Japan

Cover: Foto ©berggeist007 / pixelio.de

More available books at **www.hansebooks.com**

SUNSHINE AND SHOWERS:

THEIR INFLUENCES THROUGHOUT
CREATION.

A Compendium of Popular Meteorology.

BY

ANDREW STEINMETZ, Esq.,

OF THE MIDDLE TEMPLE, BARRISTER-AT-LAW,
AUTHOR OF 'A MANUAL OF WEATHER-CASTS,' ETC. ETC.

LONDON:

REEVE & CO., 5, HENRIETTA STREET, COVENT GARDEN.

1867.

PRINTED BY J. E. TAYLOR AND CO.,
LITTLE QUEEN STREET, LINCOLN'S INN FIELDS.

PREFACE.

———◆———

THIS work is not the account of a " vacation ramble,"
but still something that may be considered analogous.
I was led to its composition by the study of the HYGRO-
METER during the summer of last year. Every chapter
is therefore, as it were, a meditation or contemplation
radiating from that central subject, the importance of
which I have laboured to demonstrate by theory and
practice.

I need scarcely state that I have endeavoured to
exhaust the topic in all its meteorological bearings,
availing myself of all sources of information, especially
the great work of LUKE HOWARD, the Father of British
Meteorology.[1] My object has been to make the book
entertaining as well as instructive ; and it will be found

———

[1] ' The Climate of London.'

replete with practical suggestions which may be useful to all classes of readers, whilst the leading topic—the curiosities of the weather and weather-wisdom—is, I trust, explained with a fullness never before attempted, as may appear by a glance at the Table of Contents.

ANDREW STEINMETZ.

CONTENTS.

CHAPTER I.

CHAPTER IV.

CHAPTER V.

CHAPTER VI.

CHAPTER VII.

CHAPTER VIII.

WET AND DRY YEARS.

CHAPTER IX.

OF TEMPERATURE AND MOISTURE IN THE MATTER OF HEALTH IN PLANTS AND ANIMALS.

CHAPTER X.

THE WONDERS OF EVAPORATION, AND ITS SIGNS IN PROGNOSTICATING THE WEATHER.

CHAPTER XI.

THE TEMPERATURE OF THE SOIL.

CHAPTER XII.

THE EFFECT OF DRAINAGE ON THE TEMPERATURE OF SOILS, THE CLIMATE, AND RAINFALL.

CHAPTER XIII.

THE INFLUENCE OF TREES ON CLIMATE, WATER SUPPLY, AND RAINFALL.

CHAPTER XIV.

HOW TO PROGNOSTICATE THE SEASONS.

CHAPTER XV.

CHARACTERISTICS AND METEOROLOGY OF THE SEASONS.

CHAPTER XVI.

HOW TO INTERPRET THE BAROMETER AND THE THERMOMETER.

CHAPTER XVII.

POPULAR WEATHER PROGNOSTICS EXPLAINED.

CHAPTER XVIII.

THE INFLUENCE OF THE WEATHER ON WOOD, AND HOW TO PREVENT IT.

CHAPTER XIX.

THE SUPPOSED INFLUENCE OF THE SAINTS ON THE WEATHER; PERIODICITY OF CHOLERA AND ITS ATMOSPHERIC CAUSE.

CHAPTER XX.

THE CURIOSITIES OF LIGHTNING.

CHAPTER XXI.

CHAPTER XXII.

CHAPTER XXIII.

CHAPTER XXIV.

CHAPTER XXV.

WHAT BECOMES OF THE SHOWERS?

SUNSHINE AND SHOWERS.

CHAPTER I.

THE UNIVERSAL BELIEF IN THE POSSIBILITY OF
PREDICTING THE WEATHER.

THE persistence of public opinion in believing in the
possibility of a foreknowledge of the weather, is a
strange and characteristic phenomenon. Whilst a multitude of industrial and scientific ideas have had the
privilege of occupying the minds of the masses only for
a time, and then have disappeared, utterly forgotten,
even by their warmest supporters,—whilst innumerable
prejudices, strongly rooted in human nature, have been
forced to give way to the repeated attacks of science,—
the belief in the possibility of a foreknowledge of the
weather has resisted everything. From the remotest
ages,—from the epoch when man bowed down before
the stars, and adored them as the creators and protectors of our world,—from the moment when, attributing
to the heavenly bodies a determining influence on terrestrial events, he invented astrology,—the belief in the
possibility of predicting the weather has existed, and

B

survived even its numberless mistakes and failures,—in spite of the checks of those who have endeavoured to resolve this problem, in spite of the persistent resistance of modern science, which, ever since its formation, has always—down to the last few years—opposed the theory of forecasting the weather, by endeavouring to demonstrate its impossibility.

The persistence of public opinion in clinging to this belief was not only apparent in the Roman world of old, —that is, in the few regions in which all the ideas of ancient civilization circulated, and which, in their multifarious contact, must have received certain influences, certain leanings tending to keep all the world within the same circle of ideas and theories,—but we meet with it in every part of the universe, amongst savages and civilized nations. Everywhere, in the palaces of the rich, as in the huts of the poor; in the farm-house of the husbandman, as in the wilds of the savage; in the wigwam of the North American Indian, as in the dwellings of the Japanese and the Chinese, — everywhere man believes in the possibility of forecasting the weather; everywhere he receives, with marked favour, any attempt of the kind, totally regardless of the means employed.

There is, however, a great difference between such belief among civilized nations and that of savages in this respect. Whilst the former admit generally that man has no influence on the production of atmospheric phenomena,—whilst the advance of enlightenment constantly tends to suppress the belief in the supernatural pretensions of weather-wizards, and shows that man is exposed to the accidents of the elements, and cannot himself produce them,—the inhabitants of certain uncivilized countries believe in the influence of the human will on the manifestations of the forces of nature.

However, whether the belief is that the event comes to pass because it is predicted, or that it is predicted because it is to happen, the result is the same; all the world believes in the possibility of predicting the weather. It is the one universal faith of all mankind. And, if general opinion be not a scientific proof, still it demonstrates in man the existence of an instinctive sentiment, which is favourable to this interesting question. Without the least investigation, man favourably receives all that tends to support this belief. He knows that sometimes the predicted event has very little probability in its favour; but without troubling himself with this circumstance, he accepts everything,—always ready to bury in oblivion the unfulfilled prediction, and to announce, with acclamation, that which comes to pass with more or less completeness.

Current facts testify to this averment. In all the weather-almanacs, predictions are ventured for almost every day in the year, and, of course, for every part of the United Kingdom; and yet it is quite impossible that such predictions can be everywhere accomplished, excepting on rare occasions. It may be very fine weather in London, whilst " pouring in torrents" some twenty or thirty miles off, or even less. Nevertheless, these almanacs " sell well," and are highly prized by a large class of the community. A very small percentage of " hits " suffices to secure the continuance of their popularity; and if any of them should happen to make such a " hit " as Murphy made in his famous almanac for 1838, there can be no doubt that it would be attended with the same astounding result—a tremendous rush upon the publishers, with utter inability to supply the demand, and some £7000 cleared, as by Murphy on that occasion, by having—as chance would have it— marked the 20th of January as "*Fair. Probable lowest degree of winter temperature,*" which occurred to the very letter—the thermometer below zero for some hours, and a rapid change of nearly fifty-six degrees,—two things almost unprecedented in the annals of meteorological observation in this country !

Thus, a chance success intensifies the belief, and no amount of failure produces incredulity. So that, if it be true that the sentiment which we all have in us respecting the perpetuity of after-life, and the horror inspired to all of us by the bare idea of annihilation, may be taken as the best proof of the soul's immortality,— the perpetuity and the universality of the belief in the possibility of predicting the weather may be taken as a proof of that possibility.

To doubt that a science of weather is possible would be to doubt that atmospheric disturbances are governed by fixed laws. But, indeed, a wonderful change has taken place in this matter of late years. Formerly, most *savants* scoffed at the idea of predicting the weather, and Arago, the French astronomer, said that no scientific man, anxious for his reputation, would venture upon such a thing, even for the space of a single day; but now it is just the reverse; those most acquainted with meteorology are the staunchest believers in the ultimate probability of "doing the thing" to the world's great satisfaction. Indeed, had one-half of the time and research devoted to sidereal astronomy been spent in observing and registering changes, which any one may notice, but no one has yet succeeded in pre-

dicting or interpreting, meteorology would not be still
" in its infancy " after a birth thousands of years ago—
nay, coeval with the first appearance of man upon earth,
to observe the " signs which are in the firmament of the
heavens."

We have said that the failures of weather predictions
have had no influence on the minds of the masses; but
we must add that, hand-in-hand with these failures,
there are positive facts which favour and justify the
popular belief in this attribute of man. We all know
the certainty with which sailors and most country
people foresee changes in the weather twenty-four hours
before they occur, and that, too, without being able to
give very precise reasons for their prophecy. It is by a
sort of intuition that they seize, as it were, the precursors of the coming phenomena, and, for the most part,
they merely obey an instinct which compels them to
prognosticate the impending shock of the gaseous ocean
which surrounds us.

Moreover, the certainty with which the lower animals
foresee the changes of the weather, the impressions which
they receive long before man himself is influenced,
that solemn expectancy of all nature, which manifests
itself before every shock of the elements,—all tend to
show us that the forecasting of the weather is not a
mere chimera ; and that if we are often agitated, as it
were, in a circle without an outlet, if we strive in vain
to pierce the veil of the future; this happens, for the
most part, because, in our social life, there are all sorts
of concerns that prevent us from hearing the voice of
nature, and which render us deaf and blind in the presence of the most evident signs of coming weather.

Thus, then, if the lower animals get an idea of

coming weather from their sensations, man must, for
the most part, rely on those instruments which his intellect has devised, and which are, so to speak, substitutes for the instinct of the former. No single weather-
instrument is absolutely sufficient in itself, even for
ordinary indications, unless it be the HYGROMETER;
but all-sufficient as this admirable instrument is in its
capability of indicating the continuance of sunshine or
coming showers, still it is by the help of the laws established by the barometer that we are enabled to interpret its all but infallible indications. Armed with the
hygrometer, no farmer will ever be surprised by
showers, and it will enable him to count upon the continuance of sunshine on almost every occasion of such
indications. We say almost, for there are occasions,
rare, it must be admitted, when the hygrometer fails to
announce the coming change; but this unavoidable de-
fect may be easily supplied by what should always be its
companion instrument—the EVAPORATION-GAUGE, as
will appear in the sequel.

These two instruments, as we shall endeavour to demonstrate, will enable everybody to be weather-wise, at
a very trifling expense, and with very little trouble.

For our own part, we may state that since we have
set up these instruments, and studied them, we have
never been "caught in a shower" without an umbrella;
nor have we had to be inconvenienced by carrying an
umbrella without having to use it. This is rather important. For a good umbrella costs more than a good
hygrometer, such as we have been using; and it is quite
certain that umbrellas are more injured by carrying than
by use as a shelter from showers.

Of course, to the farmer, at haymaking-time and

during harvest, the use of such a monitor will be invaluable if he can be enabled to interpret accurately its indications, as we hope to enable him to do, by furnishing him with the rules of the instrument, suggested by its nature, and verified by us during a considerable experience of such appearances of the skies as were calculated to mislead, not only ordinary weather-wise people, but even the scientific " readers " of the barometer.

In the Report of the Committee of the Meteorological Department of the Board of Trade, the weather forecasts of the late Admiral FitzRoy are very severely handled, on account of their preponderant failings; and the Committee venture to decide that our present knowledge is not sufficient to enable us to predict weather; that, in fact, there is no scientific basis on which to rest daily forecasts. We totally dissent from this opinion, and have reason to believe that sufficient data exist to enable us to forecast weather, with reasonable accuracy, not only for a month beforehand, but for a season; and that, from day to day, the hygrometer may be made an infallible means of warning the farmer, the sailor, and all who may care about coming weather. If the late Admiral FitzRoy and his successors had been furnished with the hygrometric data of each station, instead of the vague initials in the column headed " weather," in their daily table, and had studied the indications of the instrument, they would have seldom been wrong in their " probabilities," and would have been spared the unscrupulous and, to a great extent, unmerited lashes of this precious committee, who, however, may be excused on the ground that they recommend the Government to vote a much larger sum than hitherto allowed for meteorological purposes, and make some valuable suggesions for the advancement of meteorology in all its bearings.

Such is our confidence in the hygrometer, after ample experience; and we promise to make good every assertion in its favour.

The simple rules which we have drawn up for guidance will be found sufficient in themselves; but it may be advisable, in the first instance, to consider summarily the physical facts and laws upon which the hygrometer depends for its indications of coming weather.

CHAPTER II.

THE ARRANGEMENT OF THE ATMOSPHERE.

THE atmosphere is a gaseous mass which surrounds the earth on all sides like an ocean, in which all plants and animals live and have their being, like fishes in the sea, which is their element, just as the air is ours. All the phenomena with which meteorology is concerned take place in the atmosphere, among which are sunshine and showers, and many others, at which we must glance in order to give completeness to our labour.

The air is said to be invisible, but this is true only respecting the limited portions immediately around us; aloft, in large masses, it assumes the blue colour which is called cerulean, resulting, however, from the watery vapour in it, and not from its essential properties. It is also said to be scentless; but no doubt, like everything else in the material world, it has some sort of scent of its own, to which, by habit, we get too much accustomed to distinguish it, and so we become sensible of its presence only by the foreign scents, good, bad, and indifferent, with which it gets readily impregnated.

As various atmospheric agents play a part in the phenomena of sunshine and showers, we must enumerate them succinctly, call to mind their chief properties, and point out the different action which they exert on one another.

The atmosphere especially destined to support the existence of the living beings on our globe consists chiefly of air, properly so called, hence usually termed atmospheric air, with which are mixed a small quantity of carbonic acid and a variable portion of watery vapour. The carbonic acid supports vegetation,—makes wood, and therefore, so to speak, made coal, which was once wood, growing in mighty forests on the surface of this beautiful world " in the far backward and abysm of time." But sunshine is necessary to enable plants and trees to make wood out of the carbon of carbonic acid ; and so, after all, it was the sun, "the Arch-Chymic Sun," as Dante calls it, that made the coal, which does so much herculean work for us. " What moves this carriage ?" asked George Stephenson, in company with an eminent *savant* in the train on one occasion. " Why, the engine, of course," was the reply. " But what moves the engine ?"—" Steam—coal."—" No, no !" exclaimed the truly philosophic British workman, " it's neither steam nor coal, but the *sun*, that made the coal, and is now working through it for ever and ever."

Various other substances besides carbonic acid and watery vapour are perpetually rising into the air, in the gaseous state at ordinary temperatures from the surface of the earth, and we are apt to imagine that they must exist in large quantities in the atmosphere ; but, on the contrary, their quantity is so small that they almost always escape notice in chemical investigations.

Gravitation, that tendency of all bodies to fall towards the earth's centre, acts upon the air as on everything else in the universe. In other words, each particle or molecule of air exerts a pressure on those which are below it. This pressure, which is thus added to the pressure proper to each molecule below it, combined with the action of the solid mass of the earth, retains the atmosphere *in equilibrio* or balanced round about the earth, and enables her to drag the atmosphere along with her in her annual movement round the sun.

If we pile one upon the other, blocks of any substance capable of being squeezed into a smaller space, such as butter, for instance, we find that every block is more and more compressed, the lowest most of all. The same *quantity* of matter exists, only it is pressed closer, its particles are nearer to each other; in other words, we give the matter greater *density*, which means that there are more particles of matter in a given space than before the pressure.

The same thing happens with the atmosphere, owing to the pressure of gravitation just explained. In a vertical or upright column of air, the densest layers touch the soil; the density then goes on diminishing in proportion to the elevation of the layers, simply because the portion of the column below the layer in question no longer exerts any pressure on the latter.

On the other hand, air being a perfectly elastic fluid, that is, always tending to resume its dimensions, the instant that pressure is removed or lessened dilates itself more and more, just as the pressure is diminished,—so that we might imagine that the atmosphere extends to an immense distance from the surface of the earth. But this is not the case. In the first place, it is evident that the expansion of the molecules or particles of air cannot be indefinite, for in that case they would spread in the celestial spaces, and go and group themselves round about the different bodies therein, to our very great loss and bereavement. Secondly, the observation of various optical phenomena, especially that of twilight, demonstrates that the height of the atmosphere does not exceed sixty miles, at which limit the tenuity or thinness of the air would be almost infinite, and incapable of refracting or bending the rays of the sun when below the horizon at the twilight hour in the morning and the evening. The following figure explains the phenomenon of twilight, which is caused by the refraction or bending of the sun's rays by the atmosphere, so as to give some light to a place situated at the proper angle to receive it..

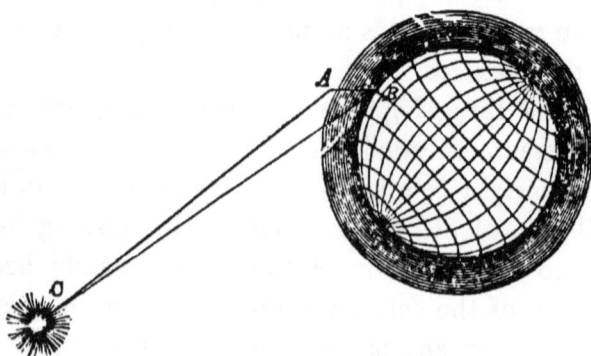

Fig. 1.

The dotted space represents the thickness of the atmosphere round the earth ; A is the point where the rays of the sun below the horizon strike the extreme edge of the atmosphere, whence it is bent in the direction from A to B, as twilight. When the sun is at any point below the horizon, it cannot be directly seen ; but, because the sun's rays can pass to that point of the atmosphere above the spectator's head, this part will be illuminated before the sun rises, or after it sets, and be visible by reflection, that is, twilight, which means *second* light, will be produced; and it is easy, by calculation from the the angle A and the sides A C and B C of the triangle, to find the side A B, or the thickness of the atmosphere, as before given; but other observations tend to show that it cannot be more than forty-five miles. For the *practical* purposes of meteorological phenomena, however, it may be taken at even half that extent; and we may roundly compare its thickness, relatively to that of the earth, to that of the *thinnest* rind of a good orange, which shows that the atmosphere is a very thin garment indeed to our planet.

We now come to the special points of the subject, which concern us in the prognostication of the weather.

The water which is spread out in such large quantities on the surface of the earth, and which circulates in its interior through an infinity of channels, evaporating at all temperatures, even when in its solid state, continually pours into the atmosphere a quantity of vapour which is greater in proportion to the temperature,—the higher the temperature the greater the evaporation, as has been proved by numerous experiments, and, indeed, as must be perfectly obvious.

Now, the altitude or height to which the vapour of water spread out in the air ascends is greater in proportion to the elevation of temperature, the greater the temperature the higher the vapour rises; and the quantity contained in the air at the same instant is also greater in the same proportion, the higher the temperature of the air the more vapour it can contain.

The second property of the atmosphere in direct connection with our subject is its electricity. The

electric fluid always occurs in the atmosphere in notable
quantities, manifesting its presence by numerous phenomena.   The storm-clouds are always charged with more
or less electricity, and rain itself is often electrical.

At a time not far distant, when no other cause of
electricity but friction was known, this cause was admitted as the source of the electricity spread out in the
atmosphere.   It was supposed to be produced solely by
the friction of the masses of air one against the other.
At the present time many scientific men totally reject
this cause; but one of them, Kaemtz, observes that
when we shake a tissue of silk in the air, it becomes
electrical; whence he concludes that the same must
result in the shaking of two masses of air of different
temperature and different humidity, as many are in the
atmosphere.   The rustling of a silk dress in our hearing
is certainly rather electrical on certain occasions, but
we don't exactly perceive, at the first blush, the connection between the two phenomena—masses of air and
rustling silk.   Silk, however, is an electric, and so is
air; but both of them require the presence of a contrary
electric for the development of their phenomena.   Thus,
if the masses of air be identical, there will be no development of electricity, just as when we rub together two
sticks of sealing-wax perfectly identical; but if one is
only warmer than the other, on the instant the colder
assumes positive electricity, and the other becomes
negative.   Now this law applies to all bodies of the
same nature when rubbed one against the other.   Hence
we may draw the conclusion that the upper masses of
the atmosphere are *positive*, whilst the lower are *negative*.   But the greatest source of atmospheric electricity
most assuredly results from the chemical actions and

reactions which are constantly going on below on the earth's surface and in the mass of the air itself. Among these, the evaporation of water, which takes place all over the earth's surface, must be placed in the first line. The experiments of Pouillet have demonstrated that the evaporation of pure water does not produce electricity; but that when we add a small quantity of salt, acid, or other soluble substances, to it, the vapour instantly becomes electrical.

This vapour is always *positive*, and the vessel whence it escapes remains *negative*. In nature water always contains some salt in solution; the vapours resulting therefore rise charged with positive electricity, whilst the soil becomes negative.

In the various acts of combustion or burning, in the respiration of animals, and the germination of plants, the carbonic acid produced is always positively electric. It is surmised that the watery particles with which the atmosphere is charged acquire positive electricity as they are rubbed by the wind against the earth and all it sustains, such as hills, rocks, trees, etc., in the same manner as the stream of steam and water become positive by rubbing against the jet. If so, what connection may not be traced between the hurricane winds of the tropics and the prevailing lightning-storms with which those regions abound?[1]

These facts demonstrate that a certain quantity of positive electricity being constantly poured into the atmosphere, this electricity must accumulate in the upper regions, which is, moreover, shown by the phenomena of thunderstorms.

The next item of our subject is *Light*, which always

---

[1] Cabinet Cyclop., "Electricity."

gives rise to heat, emanating from the heavenly bodies, and principally from the sun, continually traversing the atmosphere and producing in it a multitude of phenomena, the greater part of which belong to the domain of Optics, but many of which directly concern meteorology. Let us glance at the chief of them.

The evaporation of water, which originates the showers and all the aqueous phenomena of meteorology, results from the action of the sun's heat on all the moist parts of the earth's surface, including living beings, plants, and animals.

In the morning, after sunrise, the surface of the earth, warming under the influence of the sun's rays, communicates its increase of temperature to the layer of air which touches it. The molecules or particles of this air, becoming dilated, ascend and are immediately replaced by others which are colder, and which, after being warmed in like manner, rise in their turn; so that, during the entire day, an ascending current is thus established, or, more accurately speaking, there are currents ascending from the surface of the earth towards the upper regions of the atmosphere, and currents descending from the latter towards the surface of the earth.

The intertropical zone, being constantly more heated than the others by the sun, sends forth a great ascending current of hot air, which causes the advance of two currents of cold air from the poles towards the equator, as under the influence of a tremendous *draught*, the very tiny miniature of which we may observe any day at our keyholes, doors, and windows, cold air rushing towards warm air, to effect an equilibrium.

These effects result in the prevailing wind-currents, which are set in motion by a part of the 230,000,000th part of the sun's radiation into space, the quantity of "sun-force," as Mr. Lockyer terms it,[1] by which *all the world's work is done*, and which represents a power, a motion, or mode of motion, namely, a *heat* which would daily raise 7513 cubic *miles* of water from the freezing to the boiling point.

But greater marvels, certainly more varied and complicated effects, must be ascribed to the *radiation from the earth*, combined with the existence of aqueous vapours in our atmosphere. As the violet clothes itself in its perfume, as the peach powders itself with its bloom, so does the earth clothe itself with aqueous vapour, which is the great cloud-mother, the mystic veil of her beauty, enhancing all her charms. This, acted upon by compression and expansion, (the effect being modified in a thousand ways by the terrestrial conformation, different electrical conditions, and possibly by the varying position of our satellite, the moon,) does all the rest, as the multitudinous meteorological phenomena abundantly disclose. The heat given by a mass of vapour sufficient to give a gallon of water, when compressed, first into cloud and then into hail or snow, is sufficient to raise 67,690 gallons of air from the temperature of melting ice to summer heat.[2]

Again, a rainfall of only *one* inch (and half an inch is a very moderate amount for a single thunder-shower) over Great Britain liberates as much *heat* as would be generated by the combustion of *three hundred and fifty millions of tons of coal!* These may be taken as fair specimens of the sort of facts with which we are now familiar regarding the movements and changes of our atmosphere.[3]

---

[1] Macmillan's Mag., Aug. 1866, "Prospects of Weather-Science."
[2] Maury, 'Physical Geography.'   [3] Lockyer, *ubi supra.*

CHAPTER III.

THE MOISTURE IN THE AIR, OR HYGROMETRY.[1]

THE amount of watery vapour in the air at all times is constantly varying, and that amount has a direct bearing on the weather, or that which is coming on. In fact, the presence of aqueous vapour in the atmosphere is certainly the chief, if not the only cause of the alternations of sunshine and showers with which it presents us, as well as of all the phenomena by which these are accompanied. If the air were perfectly dry, the terrestrial surface would enjoy perpetual sunshine with gentle breezes, resulting from the general movements which the variations of the diurnal and annual intensity of the sun's heat determine in the masses of the air.

It is not always that we can detect the moisture of the air by our senses; but numerous facts will, at any time, show it at once, if we contemplate them with the enlightened eye of science; for instance, the moisture which, in hot weather especially, seems collected outside a glass or other smooth vessel, upon suddenly pouring in cold water. The reason is that by the greater degree

[1] Hygrometry, from two Greek words, meaning the measurement of moisture or humidity.

of cold in the water, the invisible vapour in the adjacent air is at once condensed on the surface, and becomes immediately visible as water. The same is also observed to happen on our windows in cold weather, from the condensation of the watery particles interspersed in the air within the room, by contact with the panes of glass, made cooler by the outer air. If the air be frosty, the glass is soon crusted over with ice; but if the weather be more open, then the water gutters down in drops. Very curious consequences may ensue from this cause on certain occasions, and in certain climates.

In Russia, during the severity of winter, if the window of a crowded ball-room be suddenly opened, the inmates are all at once startled by a *fall of snow* in the room, covering them with its flakes, just as though they were out of doors. The cause is the sudden condensation and freezing of the vapour in the air in the ball-room.

Other facts of common experience are equally illustrative of this phenomenon. There is moisture in our breath, but it is quite invisible in summer-time, being then easily received into and incorporated with the warm air around; but it is otherwise in winter, when our breath seems like puffs of tobacco-smoke as it proceeds from the mouth. The reason is that the cold air finds it difficult to receive the moisture in our breath, and so the latter remains visible for a time, proportionate to the coldness of the air into which it is breathed. Almost everything around us, if narrowly observed, will more or less testify to the presence of this vapour in the air. In "damp weather," before rain, wood "swells" with this invisible vapour, salt gets moist, catgut snaps, by contracting or shortening with the moisture, and our banisters and furniture need wiping.

We say that the air is humid or moist, when the quantity of aqueous vapour it contains becomes perceptible to our senses. It then approaches more or less the point of *saturation*, that is, the utmost quantity of vapour it can contain in suspension.

It is not, however, the absolute quantity of vapour it contains that makes the air moist. In winter the air is often moist with a smaller portion of aqueous vapour than it contains during hot dry days in summer. The reason is because the humidity of the air depends not only on the amount of vapour it contains, but also on its *degree of temperature*.

From a series of observations made at Halle, near the seashore, and on the Alps, Kaemtz established the following laws of hygrometry :[1]—

During the entire year, it is in the morning, a little before sunrise, that the quantity of aqueous vapour contained in the atmosphere attains its *minimum*. At the same time, however, on account of the maximum lowering in the temperature during the night, the humidity is at its *maximum*.

In proportion as the sun mounts above the horizon, evaporation increases on the surface of the earth, and the air receives at each instant a greater quantity of vapour. But as the temperature increases at the same time, and as the air opposes the formation of this vapour, it continually recedes more and more from the point of saturation, and the relative humidity constantly diminishes.

This decline of moisture continues up to the moment when the temperature attains its maximum ; then the humidity increases until sunrise.

[1] Meteorology.

In winter, the quantity of vapour augments regularly until towards the afternoon. When the thermometer begins to fall, the vapour condenses in part on cold bodies, and its mass diminishes until the following morning ; whilst, by reason of this lowering of the temperature, the air becomes successively moister.

It is otherwise in summer ; then the absolute quantity of vapour increases, in like manner, from sunrise ; but, before midday, there is a maximum, which, in the different months, occurs sometimes sooner, sometimes later. The quantity of vapour then diminishes up to the moment of the highest temperature of the day, without, however, attaining as low a minimum as that of the morning. It is evident that in this interval, the air always recedes more and more from the point of saturation. After having attained its minimum, the quantity of vapour again increases pretty regularly, until the following morning, whereas the air becomes relatively moister.

The march of a day's hygrometry will exemplify these rules of the air's humidity. We will select the 28th of the last July, the day before the first rain, after the long spell of north-easterly weather and drought, which occurred in that month, during the whole of which period the humidity was pretty high, often 75 and 80, complete saturation being taken at 100.

On the morning of the 28th, at eight A.M., the wind having shifted from N.W. to S.E., the humidity was 77, the temperature being 64°. At ten o'clock the temperature rose to 67°, and the humidity decreased to 68. At eleven, the temperature fell to 66°, and the humidity rose to 73. At twelve o'clock the temperature rose to 67°, and the humidity decreased to 68. At one, the temperature rose to 68°, but the humidity was still 68 ; and thus it continued from hour to hour, up to 4.30 P.M., when the temperature was 67°, as it had been at two P.M. ; and, at our last observation, at eight P.M., the temperature was 64°, and the humidity 77,—both of them exactly as at eight A.M. During the night the wind backed to the northward and westward, and the first rain fell, beginning the long course of unsettled weather which followed.

The results may differ at certain points of the globe ; and it appears that the quantity of vapour existing towards noon is relatively less marked on the seashore. Kaemst says that in order to appreciate these differences, we should be acquainted with the laws which govern the variations ; but the number of the observations being very limited, the only conclusion that may be drawn is, that all these differences depend, either upon the ascending currents of the air, or on the resistance which the air presents to the translation of the vapours. When, in the morning, evaporation begins with the increase of temperature, the vapour, by reason of the air's resistance, accumulates on the surface of the soil, as shown by observations made at every point of the globe.

This layer of vapour does not acquire great thickness ; but as soon as the ascending current begins, especially during summer, the vapours are drawn towards the upper regions of the atmosphere, with a force which goes on constantly increasing towards noon. The evaporation of the soil is then more active, on account of the increase of temperature ; the ascending current, nevertheless, still carries off the greater part of it, and there is a diminution in the quantity of vapour. This fact we have frequently verified. Towards evening, when the temperature begins to get lower, the ascending current diminishes in intensity or ceases entirely. Then the vapour not only accumulates in the lower regions of the air, but it also descends from the upper regions. Hence, it may be observed that, towards evening, there is a second maximum, which is, however, only temporary, because, during the night, when the vapour is precipitated in dew or hoar-frost, the air becomes successively drier.

The truth of these remarks is attested by another careful observer, M. Rozet, who states that on the Righi, situated above the Lake of Zurich, the midday minimum fails entirely, on account of the vapours which continually rise from its waters. Towards night, the vapours descend very rapidly below the summit of the mountain. On the Faulhorn, the greatest tension of the vapour occurs only some hours after noon.

From these notable differences in the variation of the absolute quantity of aqueous vapour, still greater differences result in that of the relative humidity. The ascending current of the morning lifting the vapours towards the upper regions, the air becomes relatively drier than it would be by reason of the increase of temperature only. Whilst these vapours rise the dryness increases less rapidly, especially if we consider that in the upper layers of the atmosphere the temperature changes less than in the lower; it may even happen, that at very elevated points the air becomes moister during the day, whilst the dryness increases towards night, when the vapours descend to the plains.

As water is constantly reduced into vapour by the action of the inherent temperature of the earth, and that of the sun's heat increased or diminished according to circumstances, by the variations of atmospheric pressure, it is evident that the quantity of vapour produced in a given time must vary with the periods of the year. This has been perfectly established by Kaemst, from a great number of hygrometric observations. In January —the coldest month of the year—the quantity of vapour attains its *minimum*, whilst, at the same time, the relative humidity is at its maximum. The quantity of vapour then increases with the elevation of the sun, and attains its maximum in July,—the month in which the air is driest. After August, the quantity of water precipitated in the form of rain, dew, and hoar-frost, is much more considerable than that which passes into the state of vapour; the quantity of the latter in the atmosphere goes on, therefore, constantly decreasing, whilst, on the contrary, the humidity continues to increase, and is, in general, greater in November and December than in January. Hence, the moist cold of these two months. This order of things prevails in all countries of the northern hemisphere, where observations have been made, and even in India, where the progression of temperature is different to that of Europe.

Whilst we know that the quantity of vapour contained in the air goes on decreasing with heat, from the equator to the pole, it has not been established whether the relative humidity increases in the same relation. On the ocean, in all latitudes, the air seems to be in a state of saturation; and yet, the water of the ocean containing several salts in solution, gives off less vapour than pure water. A much colder mass of distilled water, equal to that of the ocean, would give off the same amount of vapour given off by the latter.

As air deposits a portion of its vapour on all bodies

colder than itself which it happens to touch, the point
of such deposit of moisture is called the Dew Point,
which is the temperature of cooling bodies when it has
just become sufficiently low to allow the ambient air to
deposit a portion of its humidity in the form of dew
drops. Now, on the surface of the ocean, the temperature of the dew-point is generally lower than that of
the water ; the conclusion is that the air above it must
be always completely saturated.

The quantity of vapour is greatest on the seacoasts,
in equal latitudes ; it thence diminishes in proportion as
we advance into the interior of continents.

In Algeria, at the stations on the coast, after the
driest and hottest days, immediately after sunset, the
soldier's uniform becomes wet with dew, and in a single
night the blades of knives in the pocket become rusted ;
but at thirty miles from the coast, night exposure is not
attended with any inconvenience from the humidity of
the air. This rule holds good in the interior of the
United States, the deserts of Africa and Asia, the steppes
of Siberia, and the interior of New Holland. The deserts of Africa, being entirely arid regions, present no
evaporation ; besides, the extreme heat, which is increased by the reflection of the sands, opposes aqueous
precipitation, and is one of the causes of the eternal
sterility of that country.

During a course of observations on the Alps and the
Pyrenees, carried on by M. Rozet, he made a very curious discovery, which he describes as follows :—" I have
often seen the atmosphere of aqueous vapour terminated
by a horizontal surface similar to that of the ocean.
This beautiful phenomenon only shows to advantage
and distinctly when the air is calm and tolerably pure.

In the ordinary state of the atmosphere, at all hours of the day, up to a certain height, I found myself plunged in a mass of aqueous vapour more or less dense, according to the atmospheric circumstances, sometimes hiding from me the signals placed at a distance of only about seven miles from my observatory, and on other occasions permitting me to see those which were about twenty-seven miles distant. Whilst ascending the mountains I remained in that mass of vapour up to an altitude more or less considerable, according to the degree of temperature; but afterwards the atmosphere became much clearer, and I beheld the *upper limit of the atmosphere of vapour traced on the horizon by a bluish line,* similar to that which bounds the horizon of the ocean. When I was on the northern slope of the Pyrenees, this line formed before me a semi-circumference, having for its diameter the axis itself of the mountain-chain; and often, when I was on summits whence I could see at the same time both France and Spain, it was a complete circumference of which I was myself the centre. Very little above the level of the blue line, in calm weather and with a pure sky, I saw at its level, the atmosphere of vapour bounded by a perfectly horizontal surface, in the Pyrenees as well as in the Alps, starting from the chain and terminating at the same line. Towards noon, when the sky was very pure and the air calm, the terminal surface was formed by a light fog across which I beheld the ground situated below me, as though covered with a light gauze, and the summits above me seemed like islands in the midst of a vast ocean. Continuing to ascend, I saw the nearest portions of this foggy surface disappear,—that is, there seemed to be formed in the central part of the circle a void or open space which

increased in extent the further I mounted; and to that
extent, that on summits about 900 feet high, I frequently only perceived a narrow zone near the horizon,
and concentric with the bluish line. When plunged in
the atmosphere of vapour it seemed bluish at the side
towards the sun and whitish on the opposite side; when
I got higher, and the sun was rather high above the
horizon, the upper surface presented here and there certain different tints, which were scarcely perceptible; but
in the evening and in the morning, towards sunset and
sunrise, it was sometimes covered with a beautiful pur-
plish light, forming a spherical streak whose thickness
varied in the inverse ratio of the sun's altitude above the
horizon." [1]

This is an important fact in meteorology, pointing to
a definite cloud-area in the skies. From M. Rozet's observations, it appears that the regularity of the terminal
surface of this vapour-atmosphere persists during the
entire day, when the weather is fine and the air calm
or nearly so. The altitude of this surface is not the
same for the different hours of the day; it increases
and diminishes with the altitude of the sun, and during
different days, if the weather continues fine, it is higher
in proportion to the elevation of temperature. Saussure
observed the same fact in his Alpine travels, and found
that the terminal surface of the vapour-atmosphere is
composed of aqueous vapour.

This level of the terminal surface of the vapour-
atmosphere, at any hour of the day during fine weather,
is the region where the vapour of water, which, by
reason of the temperature of the earth's surface, continually rises in the atmosphere, passes from the *invisible*

[1] De la Pluie en Europe.

state into the visible or *vesicular* condition. At that
level are arrested the greater portion of the aqueous
vapours produced on the earth's surface; and it is evident, that when there is an equilibrium (in other words,
during fine weather) there must be, comparatively, a
great degree of *dryness in the regions above it*. But
when clouds descend from the upper regions of the
atmosphere at the moment when others ascend from the
lower, to accumulate on the summits of the mountains,
in stormy weather, it is obvious that there must be a
greater degree of humidity on the elevated points than
in the plains and valleys.

The hygrometry of the Arctic regions is very curious.
Although the air in the Arctic seas is generally in a
state of dampness, approaching to complete saturation,
yet the absolute quantity of moisture cannot be considerable when the cold is very excessive, as before
explained. It is remarked, that vessels are less apt to
rust here than in any other climate, and this observation,
if we consider the relative humidity of the atmosphere,
as indicated by the hygrometer, is certainly correct.
But though the air in the Polar regions is generally
damp, yet it is probable that there is no habitable situation in the known world, in which such a degree of
actual dryness prevails, as in a house or in the cabin of
a ship well heated, whilst the external air is intensely
cold. The wainscoting of the cabin of a ship, in cold
weather, sometimes shrinks, in consequence of the uncommon dryness, as much as half an inch in a panel of
about fifteen inches broad, being equal to one-thirtieth
of the breadth; but on returning to Britain, the same
panel expands again to almost its original dimensions.

CHAPTER IV.

THE INFLUENCE OF THE DIFFERENT WINDS IN
THE PRODUCTION OF SUNSHINE OR SHOWERS.

THE humidity of the air varies with the different winds,
some winds being moist, some dry, and others partaking
more or less of the respective qualities. The result is a
material difference in the drying power of the various
winds, and this is rather an important matter to those
who must depend upon the winds for drying purposes,
such as farmers, for drying their corn and hay, or laun-
dresses and bleachers and dyers, for their linen stuffs,
etc. In general, provided the wind blows continuously,
their wishes are soon satisfied, but experience tells them
that a much longer time is required with a west wind.
Indeed, certain operations in dyeing cannot be success-
fully carried on without east winds, any more than
others without bright sunshine, for instance, scarlet
dyeing.[1]

Everybody is familiar with the process of drying, but

[1] Unable to account for the superiority of the French scarlet dyeing,
some of our English manufacturers consulted the French dyers, who,
upon explaining their method, were met at every stage with the remark,

perhaps very few consider the nature of the operation,
and yet it involves a most important meteorological fact,
—the capacity of the air for absorbing water; for drying
is really nothing more than the absorption of the moisture exposed to it by the substance dried,—in other
words, it is merely a process of evaporation, the wind
carrying off the vapour; and the amount of evaporation
has a definite relation to the wind that blows.

The range of evaporation corresponding with the
points of the compass is as follows :—It is least when the
wind blows between N. and N.E.; it increases when the
wind shifts to E., S.E., and S.; and reaches its *maximum* between S. and S.W., to diminish again in passing
to W. and N.W.

The cause of these different effects becomes apparent
from a little consideration. Before arriving at our
shores, the west winds pass over the Atlantic, and become charged with vapour, whereas those which blow
from the east come from the interior of the continents
of Europe or Asia, and having passed over land, they
have parted with their moisture and come to us dry,
arid, and "parching" with thirst, and so they readily
imbibe moisture from the air, the clouds, and everything
they touch, putting an end to showers if they do not
always secure sunshine.

As every wind with *west* in it is more or less connected with that which comes laden with moisture from
the Atlantic, it will not be surprising to know that

"Exactly what we do," until at length the Frenchmen said, "Finally,
you must select a fine, sunshiny day for the operation."—"Ah! there's
the rub! We can never *select* such a day in England, and so you will
always beat us at scarlet," was the unfortunate conclusion of the depu-
tation.

westerly winds have the greatest influence on the degree of humidity in the air.

Now England is certainly a moist country, and it results from the fact that our southerly and westerly winds are to our northerly and easterly winds, as 196·4 is to 135. In a mean of years, we have the following winds :—

North	40 7 days.
North-east	47·6 ,,
East	22·6 ,,
South-east	19·9 ,,
South	34·2 ,,
South-west	104 ,,
West	38·3 ,,
North-west	24·1 ,,

These winds have different tendencies to "hang" or last, and, according to the Astronomer Royal, there are only *eight* points from which the wind ever blows steadily for any lengthened period in England. He says, that it never blows at all directly from due S., and that the two most prevalent winds are the S.S.W. and W.S.W., the former invariably bringing rain, and the W.S.W. accompanied by dry weather,—that is, of course, comparatively dry, a few days of sunshine with barely a chance of showers. Between W. and N.W., there is another point of hanging or duration; also between N. and E., the former bringing unsettled weather, the latter a spell of dry weather; the next point of duration is between E. and S.S.W., with unsettled weather; and finally, we have durations with N., W., and E. winds, which make up the eight points suggested by the Astronomer Royal. But N., W., and E. may be considered "ticklish" points, as our winds seem to like to get married as soon as possible, so we seldom find them in the above state of single blessedness, especially as they have not far to go to find a wife, on the right or left of the compass.

The relative quantity of vapour for each wind, as given by Kaemst, from experiments at Halle, are as follows, and they may be taken as sufficient approximations in England, according to our observations :—

North wind . . . . 78·3
North-east . . . . 77·5
East . . . . 73·0
South-east . . . . 74·8
South . . . . 73·6
South-west . . . . 74·8
West . . . . 74·4
North-west . . . . 76·5

Thus, as Kaemst observes, although during the north wind the air contains a much less proportion of the vapour of water than during a south wind, it is nevertheless infinitely more *moist*, on account of its low temperature. On the 29th June, 1866, after five days' continuance of N.E. wind, alternating with 80 and 70 humidity, there was half an inch of rain with the humidity at 80.

The fluctuating humidity of the *seasons* of the year, in connection with the different winds, is an important meteorological question with reference to sunshine and showers. The following observations supplied by Kaemst, as made at Halle, may serve to give some idea of the variations :—

Wind.	Winter.	Spring.	Summer.	Autumn.
N.	89·5	75·0	67·6	78·7
N.E.	91·2	72·3	67·4	82·6
E.	92·6	66·9	61·3	75·7
S.E.	85·5	71·4	66·3	79·2
S.	83·0	70·3	67·4	76·2
S.W.	81·9	70·3	69·9	78·6
W.	80·9	71·7	71·4	80·6
N.W.	83·2	73·4	68·8	82·7

Although this table may differ from the records of our seasons for the present year, it is probable that it gives the average of humidity for the different seasons with sufficient accuracy for meteorological conclusions.

The contrast between winter and summer is striking. Although in these two seasons the proportion of vapour is less during east than during west winds, yet the lower temperature of these winds during winter re-establishes the equilibrium, and in this season the east wind is the moister, and the west the drier. In summer the contrary is the case. When either of these winds is commencing to blow, the contrast is the more striking. If, for example, in winter, the west winds have prevailed for some time, with a very pure atmosphere, and an E. or N.E. wind suddenly rises, the sky becomes cloudy in a short time; one part of the vapour of water is precipitated in the state of rain or snow, and thick fogs occupy the lower regions of the atmosphere.

In this state of things the barometer often stands at FAIR, which apparently justifies the complaint of ordinary observers respecting the false predictions of the instrument. But, if the east wind continues to blow, the sky then becomes serene, although the air remains moist. If the reverse takes place, that is, if the sky is clouded, the wind being in the east, and it suddenly passes to the south, the sky becomes clear and the atmosphere dry; because the warmer air dissolves the vapour of water and is far from the point of saturation. Hence a shift of wind to the southward will sometimes bring a fine day or two after wet. It is only when this wind has prevailed for some days, and has brought to us a large quantity of vapours from the Continent, that the atmosphere again becomes moist.

D

Vapour brought to us by a S. or S.E. wind is decomposed by our prevalent N.W., and in some cases W. and S.W. winds. This vapour must have been generated in countries lying to the south and east of our island. It is therefore probably in the extensive valleys watered by the Meuse, the Moselle, and the Rhine, if not the more distant Elbe, the Oder, and Weser, that the water rises, in the midst of sunshine, which is soon afterwards to form our clouds, and pour down in our thunder-showers. And this island, in all probability, does the same office for Ireland; nay, the eastern for the western counties of South Britain. "After nearly nine days' wet weather," says Luke Howard, "attended as usual with mixed winds, in our district, upon the wind changing from S.E. to N.E. it became fair with us; and on the same day, a rain of three days and nights commenced in the country east of the Upper Rhine about Stuttgard, so heavy as to produce a serious inundation. In the meantime we had no rain, although the barometer was still very low, and the change of the wind above mentioned had been attended with thunder. The rain ceased at Stuttgard two days afterwards, and on the next two days it rained again with us. To suppose a connection of the phenomena at this distance with electrical principles may be too much; but I think one may be made out through the medium of the winds, in the following manner. The evaporation of a tract of country lying to the east of both stations, might in the first instance be conveyed to the Thames, and then, by a change in the direction of the prevailing wind, to the sources of the Rhine; and decomposed into rain with us by the effect of a colder latitude, and with them by that of the elevation of the country, aided probably in both cases by opposing currents."

Thus drought and sunshine in one part of Europe may be as necessary to the production of a wet season in another, as it is found to be on the great scale of the continents of Africa and South America, where the plains during one half of the year are burnt up, to feed the springs of the mountains, which in their turn contribute to inundate the fertile valleys and prepare them for a luxuriant vegetation. In 1816, the middle of Europe was subjected to excessive rains, at the same time that the north, or the parts east of the Baltic, about Dantzig and Riga, were suffering from drought, and in all probability furnishing the water!

In the spring and summer, both the direction of the winds and the relative state of temperature, seem to forbid our receiving much rain from the Atlantic; but in winter, when the surface of the ocean is giving out heat to the air, it may be supposed also to give out vapour in greater quantities than the temperature of the air is prepared to sustain.

Hence, the Atlantic, during the winter months, or rather in the interval between the autumnal equinox and the winter solstice, is probably the great source of our rains. The impetuous gales, which, at this season, move over its surface and impinge on our western shores, may possibly bring us much vapour from the upper atmosphere of the tropic in which they originate. The powerful manifestations of electricity which at times attend them, seem to favour this opinion; but, should they have deposited much water on the passage, we may still find in the relative winter temperatures of the air on our coasts, and on the ocean, a sufficient reason for the *turbid* state in which they are almost uniformly found on their arrival.

A wind between north and east is always connected
with our driest season about the spring equinox, and a
wind between south and west with the wet season following the autumnal.

There is a remarkable regularity in the succession of
the winds in the first six months of the year, influencing
their sunshine and showers. They are divided into
classes as follows :—

> In January, the wind blows between W. and N.
> „ February   „   „   S. „ W.
> „ March   „   „   N. „ E.
> „ April   „   „   N. „ E.
> „ May   „   „   S. „ W.
> „ June   „   „   W. „ N.

After June the class between W. and N. prevails during
the summer, and the class between S. and W. through
the latter months of the year.

The following is a table of the annual amounts of the
several classes, as recorded by Luke Howard in a space
of ten years, together with their *amounts of rain*.

Year.	N.–E.	E.–S.	S.–W.	W.–N.	Variable.	Rain.
1807	61	34	113	114	43	20·14 in.
1808	82	38	108	103	35	23·24 „
1809	68	50	123	91	33	25·28 „
1810	81	72	78	83	41	28·07 „
1811	58	59	119	93	36	24·64 „
1812	82	66	93	91	34	27·24 „
1813	76	53	92	124	20	23·56 „
1814	96	65	91	96	17	26·07 „
1815	68	36	121	107	33	21·20 „
1816	64	66	106	102	28	32·37 „
Averages	74	54	105	100	32	25·18 in.

This table shows some very striking results as to the manner in which the several annual quantities of winds are related to those of the rain.

In the driest year of the whole, which is 1807, the class N.-E. has nearly double the number of the class E.-S.; in 1815, the next for dryness, the same; and in 1808, which stands third, rather more than double.

In 1816, the wettest year, on the contrary, the class E.-S. *exceeds* the N.-E.; in 1814, it has two-thirds of the amount of the latter; in 1812, three-fourths; and in 1810, the remaining wet year, the amount comes within a month of the N.-E., both classes being large, and the westerly winds falling off in a remarkable manner to make room for them.

The year 1811, which presents about an average of rain, has the features, in respect of rain, of a wet year. On examination, we find that thirty-six out of the fifty-nine observations here forming the E.-S. class are put down as an east wind, and in 1809 and 1813, the two remaining years, both a little below the average, the majority of the observations in the second class are of the same kind.

These proportions, then, confirm the relation of a *N.E. wind to the dry weather*; and they establish another relation between a *S.E. wind and the rain of our climate*.

With regard to westerly winds, the class W.-N., we may observe, falls off gradually during the three years following 1807, while the annual rain increases from year to year; and in four of the remaining years, its number is above the average in the dry years, and below it in the wet ones. There is, therefore, a manifest general relation of this class *W.-N. to our fair weather or sunshine*.

The winds between the south and west have no decided connection with either a wet or a dry *year*.

This statement may seem at variance with the connection exhibited in the table previously given; but the
contradiction is apparent, not real, as will be seen in
what follows, suggested by Luke Howard, supporting
his explanation with numerous examples.

There are two ways in which we may conceive rain to
be produced in a temperate latitude; first, by the cooling of the whole mass of the atmosphere to a degree
sufficient to decompose its vapour. This happens when
either the air, flowing constantly from south to north,
leaves the influence of the sun behind it, or the sun, declining in autumn and retiring to the southward, leaves
the air to cool where it remains. In effect, both causes
may be in action together, as is probably the case during
some part of every autumn in these latitudes.

Secondly, by the cooling of a portion only of the air
from the intrusion or the overflow of a warm, vaporous
current, from a lower latitude into our own, where it
loses its heat, and has its vapour decomposed by our
colder air.

In the first case, the rain will be formed in every
part of the atmosphere, up to a certain height, at least,
from the ground, where the vapour, diffused through a
rarified medium, can afford only a kind of hazy precipitation, which gradually descends upon the lower air. In
the second case, showers, and hail with thunder, if the
contact be very sudden between the currents, are generated, which fall from a greater height, and are
commonly much less *continuous* than the other kind of
rain.

Both these modes of precipitation, again, may be in

action together for a time. A *southerly* current, charged with vapour from a warmer region, may be passing *northward* at the same time that a *northerly* current may be returning towards the south, in the immediate neighbourhood of the former; and these two may raise each other, the colder running in laterally under the warmer current, and causing it to flow over laterally in its turn, while each pursues in the main its original course. In this case, the country, for a considerable space extending from about the line of their junction far into the southerly current, may be the seat of extensive and continued rain.

According to Dalton and Luke Howard, the first six months of the year must be considered as dry months, and the last six as wet months, April being the driest, and the sixth after, or October, the wettest. If this be so, and unless our seasons be altered, the present year (1866) will be excessively wet; for, by the end of June, we had by far the greater proportion of the usual rainfall all over the country, as will be more particularly considered in the sequel.

The proportions of rain in the year stand as follows, according to Luke Howard :—

For the first six months (January to June), 10·174 inch.

For the latter six months (July to December), 14·630 inch.

It appears that more rain falls during the night than by day,—the rain by day being only two-thirds of the quantity that falls by night.

In our climate, *on an average of years*, it rains nearly every other day, more or less. The number of days (of twenty-four hours), on which there falls rain, is less in the longest days than in the shortest, in the proportion of two to three. This propensity to frequent dripping may be connected with our moderate and variable temperature. In climates, the mean temperature of which, from the circumstance of latitude alone, departs further in either direction from the mean temperature of the earth, it is probable that the number of *dry* days will be found greater, in proportion as the climate is hotter or colder than our own.

The effect of the moisture in the air is everywhere apparent. In some countries, where the wind blows almost constantly in one direction (and of course the wind is only air in motion), vegetation is abundant on one side against which the wind blows from the ocean— an inexhaustible source of moisture, while on the other side there is scarcely any vegetation at all, as in Peru, Patagonia, parts of Arabia, and Africa, various islands, parts of Asia, and Australia, where all the moisture from the sea winds has been condensed and extracted; and, in passing across extensive land, the other side, as in Peru, receives only dried air from the land, and the country is more or less barren.

So in many places; wind carrying moisture affects one side of a hill or mountain, and does not affect the other equally.

Farmers should keep an eye to this fact, and always consider whether their acres lie in a good direction for securing a good supply of moisture. Here, in England, being exposed to south-westerly, westerly, and southerly, and south-easterly winds for about three-fourths of the year, we have been remarkably favoured in this respect, because these winds are not only moist, but most of them warmed in passing over the warm waters of the ocean, both the tropical and those of the more expanded or warmest parts of the Gulf Stream, which may be called the warming-pan of the air-sheets of Europe in general, and of England in particular.

Whilst air thus abundantly absorbs moisture or water, it appears, on the other hand, that water absorbs air in great quantities, and there is always a definite quantity of air in water. This capacity in water for absorbing air may be shown by an easy experiment. If water be well boiled and left to cool, it will be almost entirely deprived of the air in it; then take a glass phial of it, stop and turn it upside down, leaving the space of a hazel-nut between the cork and the water. In twenty-four hours' time the vacant space will disappear, and the phial will seem to be filled with water, the air in that space having been absorbed by the water. More air may be still added to the water, increasing its bulk proportionately; but when the water is thoroughly saturated with air, it will take no more.

Thus it appears that water is constantly preying on the air; and herein, perhaps, is the explanation of the fact that rain often stops or moderates a wind, namely, by absorbing much of it according to its condition. In general, if it rains after a high wind has set in, the wind moderates, or, as the sailor's proverb has it—

"When wind comes before rain,
Soon you may make sail again."

CHAPTER V.

THE HYGROMETER, ITS RULES AND WEATHER-INDICATIONS.[1]

OUR ancestors devised various means for indicating the increase or diminution of moisture in the air, many of them extremely ingenious, but of course not to be compared in their results to the scientific instrument of modern times, to which the name of HYGROMETER, or "measurer of moisture," has been given. First, either by suspending a little weight at the end of a well-twisted elastic string or piece of whipcord, or else by fixing such weight in the middle of the string horizontally hung loose over a couple of tacks. The weight in this case serves as an index, and a few marks made on a perpendicular scale will show the variations of the air in the extent of moisture. These effects become very evident by taking the appliance into a damp vault, and afterwards setting it near a fire.

The twisted beard of a wild-oat, with a small index fixed to it and a circular scale, will make a rough but excellent hygrometer, for it will move by the humidity

---

[1] A portion of this chapter first appeared in the 'Mark Lane Express.'

of the breath while you look at it, unless particular care be taken to prevent it.

Catgut will also serve this purpose very well, as may be observed by the coming out of the *Man* or his *Lady*, depending on the expansion or contraction of the catgut, in the Dutch toy called the *Weatherhouse*, or by the play of the cowl in the other weather-toy called the *Capuchin*.

But, indeed, almost everything around us may be consulted as rough hygrometers, for everything in nature is more or less *hygrometric*, that is, more or less sensible to moisture, owing to its porosity, or small pores invisible to the naked eye, which absorb moisture or part with it, and so expand or contract in proportion. Thus, when the air is moist, the particles of water floating abundantly in it fix on all bodies exposed to it; they enter the pores of such as are of a loose and open texture, and consequently expand or swell them very much. Hence wooden doors, drawers, and the like, are generally found to stick in moist weather. If wood be not painted, it imbibes the watery particles as fast as they form; if it be painted, they settle on the surface; and in damp weather, the wainscot is covered with a dew, and the moisture sometimes gutters down in drops. Glass, marble, and other dense bodies of a smooth surface, show this on every occasion; but cordage, string, and the looser bodies, admitting moisture in their open spaces, show the effect more sensibly.

Moisture, then, always exists in the air, in greater or less quantities, under the form of vapour, and the amount of it is accurately indicated by the hygrometer.

Organic bodies were used by the early meteorologists as hygrometers, owing to the property of such substances to absorb the moisture of the air. For instance, a hair was boiled in a weak solution of soda in order to remove the grease, one end was fixed in a frame, while the other was rolled upon a cylinder which carried an index. When placed in air saturated with moisture, the hair elongates, whereas it shortens in dry air. Such is De Saussure's hygrometer, in which he placed the needle at a point marked 100 when the air was completely saturated with moisture, and 0 corresponded with the condition when it was completely dry, the interval between the two points being divided into 100 parts; hence our modern mode of representing the moisture of the air.

De Luc's hygrometer, much used by the early meteorologists, among the rest by Luke Howard, consisted of a thin piece of whalebone cut across the fibre and fixed at one end by a gold wire to a delicate wheel carrying an index. The whalebone lengthens as it absorbs moisture, and shortens as it becomes dry, and thus acts as a hygrometer. But the degrees indicated by these instruments are not proportional to the known quantities of vapour in the air. When they point to 80, the air often contains not 80, but merely 60 or 70 per cent. of the quantity of vapour which would be necessary to saturate it. Mr. Casella has a very ingenious little *Beard of Wild-Oat* hygrometer, which, upon experiment, gave tolerable approximations to the degree of humidity in the air, as shown by the Dry and Wet Bulb hygrometer; but all such instruments will always labour under the serious disadvantage of deteriorating by time or exposure, owing to the very nature of their materials, which become less and less sensitive, and finally lose all their hygrometrical properties.

At length, Dalton suggested the use of the thermometer as a hygrometer, and his method or process was ultimately perfected by Daniell in the instrument which goes by his name.

There is another hygrometer, perhaps equally ingenious, invented by Regnault, and described by Mr. Glaisher as follows.[1] It consists of two delicate thermometers kept in position by passing through corks fitting into long cylindrical cups of polished gold or silver, one of which is partly filled with ether, so that a portion of the bulb of its thermometer is immersed in it. A small tube passes through the corks, open at both ends, to which a flexible tube is fixed of any length, allowing the observer to be as distant from the instrument as he pleases. By this means air is drawn from or driven into the cups, and the ether evaporates with a rapidity depending on the current, which is obtained at pleasure; dew is deposited on the cup containing the ether, the slightest dulling of which is seen by contrast with the other cup, which continues bright, and the dew-point ascertained. But the same objections apply to this instrument as to that of Professor Daniell, and must ever secure the preference to the very simple, easily manageable DRY AND WET BULB HYGROMETER.

The principle of this instrument, at first sometimes called *Psychrometer* (from *psychros*, "cold"), seems to have been originally suggested by Hutton, according to Kaemst; it was modified by Leslie, and then restored to its primitive simplicity by M. August of Berlin, who used it in 1825; Gay-Lussac had used it as early as 1822; but it was Dr. John A. Mason who seems to have brought the instrument into vogue in England in

[1] 'Hygrometrical Tables.'

the shape which goes by his name,—two thermometers placed side by side, with a tube containing water between them, supplying wet to one of the bulbs. Dr. Mason visited Madeira in 1834–35 for the benefit of the climate, and from the delicacy of his health he naturally turned his attention to the varying states of the atmosphere. During his residence in that island, he constructed the well-known Mason's hygrometer, and made an extensive series of experiments in every variety of circumstances, comparing it with Leslie's and Coumel's hygrometer; and on his return to this country he wrote a valuable paper, descriptive of the instrument and of his experiments, which was published by Dr. D. Thomson in 1836.[1]

The instrument gradually came into use; but it received great impulse from the labours of Mr. Glaisher, who slightly modified it by substituting a cup for the central tube, and published a set of tables for the various results obtainable by its employment.

This instrument consists of two thermometers, as before stated, placed side by side, at a convenient distance to prevent interference in their respective indications, and they should be as nearly as possible alike in every way, one being marked Dry, the other Wet,—mounted on a porcelain slab, upon which the divisions and words are engraved and permanently burnt in, so that they cannot be obliterated by any length of exposure to the weather, and may be cleaned like ordinary porcelain.

As suggested by Mr. Glaisher, the thermometers should have very small bulbs, so as to be delicate and sensitive in the extreme, the thread of mercury very

---

[1] 'Records of General Science,' vol. iv. Also in Farr's 'British Medical Almanac,' 1837.

fine, the graduations should be on their own stems, and their readings should be compared with those of a standard thermometer before use, to determine their index errors.

The bulb of one thermometer is covered with a piece of muslin, from which a few inches of lamp-cotton trail into a small cup fixed below it, containing rain-water, which ascends through the cotton and keeps the bulb moist or in a state of evaporation.

Before use, the cotton-wick, according to Mr. Glaisher, should be washed in a solution of carbonate of soda, and pressed under water throughout its length; but M. Cavalier, the well-known meteorologist of Ostend, in a note to Mr. Casella, is of a different opinion. He says:—" I have found the imbibing quality of the cotton and muslin employed with the psychrometer greatly augmented by first washing it in *diluted* sulphuric acid, and afterwards, when well rinsed in clear water, in ammonia; and further, the incrustation which I have always found to form on the bulb of the thermometer, is entirely prevented. This process, you will see, is quite different from that recommended by Mr. Glaisher, namely, carbonate of soda. The acid removes the limy property of the gum, as well as the gum itself, with which all manufactured cotton is more or less impregnated; and the ammonia necessarily destroys any greasy matter that remains."*

It is impossible to overrate the hygrometer as a means of foretelling coming weather. Indeed, the knowledge of a few facts will suffice to show that the hygrometer is indispensable if we wish to form an accurate opinion on this subject. One of these facts is, that the air is rendered lighter not only by its expansion, but also by the

* ' Proceedings of the British Meteorological Society,' vol. ii.

amount of invisible *watery vapour* in it; and this is accurately shown by the hygrometer, thus enabling us to distinguish whether a fall in the barometer indicates wind or rain, or both together.

In connection with the barometer and thermometer, this instrument affords infallible notice of impending weather—provided we know how to interpret and reason out its indications, which is by no means difficult.

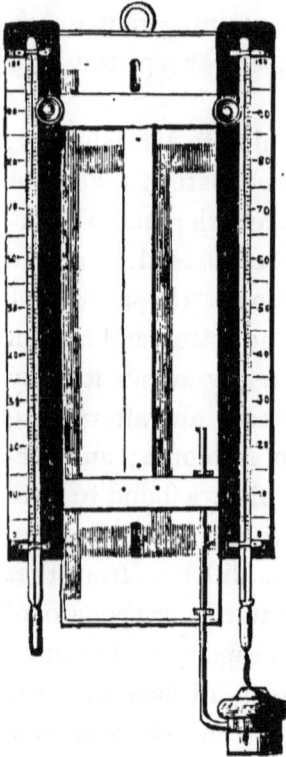

As the instrument must be used out of doors, it is obvious that the hygrometer attached to what is called "the farmer's barometer," is a very useless affair as an indicator of weather. It can only serve to show the amount of moisture in the air of the room where the barometer is hung, and that imperfectly. In testing the moisture of the air for the prognostication of the weather, the hygrometer must be suspended in a convenient position out of doors, free from draught, on a proper stand, sheltered from rain, about four feet from the ground, not very near buildings, nor exposed to sunshine. Of course the instrument may be kept in-doors, and only carried out for observation after five or ten minutes' exposure.

Fig. 2.

Wet-Bulb Hygrometer.

In summer time apartments are cooled by sprinkling water on the floor, and the same happens to our streets after the watering-cart has passed. The reason is that the water evaporates, and, in so doing, takes from the surface the amount of heat which is requisite for its passage into the state of vapour—the consequence being the production of cold. This is the way the wet-bulb thermometer acts. As the water in the cotton evaporates, it takes from the bulb that amount of heat required for its conversion into vapour, and the consequent cold is shown by the fall of the mercury in the tube.[1] If the air be absolutely saturated with moisture, which is sometimes the case, the two thermometers will read off alike; if it be very dry, then the difference between the two will be considerable.

The difference between the two thermometers doubled and taken from the lower, will approximately show the dew-point,—that is, the temperature to which air must be reduced at any time before the moisture in it, in the invisible form of vapour, can be deposited as dew. But this rough method only answers at a few temperatures; in the great majority of cases it will make it either too high or too low. Of course the nearer the approach to saturation with moisture, the greater the tendency in

[1] Our skin tells us the same when covered with perspiration; for, as the latter evaporates, we experience a sensation of cold. This evaporation is much less active in *moist* than in dry weather: hence, the sensation of cold is much greater in dry weather: hence, also, in summertime, when there is no wind, we find the heat insupportable if the air is *moist*—that is, unable to take in more vapour—although the thermometer does not show great heat; but if there be wind to remove successively the air saturated with the *vapour* surrounding our body, then the evaporation takes place with greater rapidity, and we feel cool,—the air is no longer "sultry" or "oppressive." *Fanning* effects this by bringing fresh air to the heated skin, to act as just explained.

E

the air to part with it in dew or rain by the lowering of the existing temperature.

A set of tables has been drawn up by Mr. Glaisher, which give by inspection the dew-point and percentage of moisture in the air for all temperatures of the dry and wet bulb. Complete saturation is taken at 100, and thus the other states are expressed by decreasing the numbers according to their proportion of moisture.

There is also another method of ascertaining the dew-point, namely, by calculation with a set of " factors " furnished for the purpose.

TABLE OF FACTORS FOR COMPUTING THE DEW POINT.

(*From the Greenwich Meteorological Volume for* 1857.)[1]

Reading of the Dry Bulb Thermometer.	Factor.	Reading of the Dry Bulb Thermometer.	Factor.	Reading of the Dry Bulb Thermometer.	Factor.	Reading of the Dry Bulb Thermometer.	Factor.	Reading of the Dry Bulb Thermometer.	Factor.	Reading of the Dry Bulb Thermometer.	Factor.
20°	8·1	32°	3·3	44°	2·2	56°	2·0	68°	1·8	80°	1·7
21	7·9	33	3·0	45	2·2	57	1·9	69	1·8	81	1·7
22	7·6	34	2·8	46	2·1	58	1·9	70	1·8	82	1·7
23	7·3	35	2·6	47	2·1	59	1·9	71	1·8	83	1·7
24	6·9	36	2·5	48	2·1	60	1·9	72	1·8	84	1·7
25	6·5	37	2·4	49	2·1	61	1·9	73	1·8	85	1·7
26	6·1	38	2·4	50	2·1	62	1·9	74	1·7	86	1·7
27	5·6	39	2·3	51	2·0	63	1·9	75	1·7	87	1·6
28	5·1	40	2·3	52	2·0	64	1·9	76	1·7	88	1·6
29	4·6	41	2·3	53	2·0	65	1·8	77	1·7	89	1·6
30	4·2	42	2·2	54	2·0	66	1·8	78	1·7	90	1·6
31	3·7	43	2·2	55	2·0	67	1·8	79	1·7	...	...

These " factors " must be multiplied into the difference of the two thermometers, and then the reading of

[1] Opposite each *temperature* will be found its *factor*.

the dry bulb, *minus* the product, will be the dew-point temperature.

For example :—

<pre>
Suppose the dry-bulb thermometer ‌ 63·5°
        „   wet „   „   57·3
                                   ─────
Difference . . . .   6·2
Factor (from the Table) . . 1·9
                            ─────
                             5·58
                             6·2
                            ─────
                            11·78
Dry-bulb reading . . . . 63·5
Temperature of dew-point . . 51·72°
</pre>

By means of the dew-point, accurately ascertained, many points of the utmost interest to chemical and meteorological science may be determined. By mere inspection of tables properly constructed, such as those published by Mr. Glaisher, we can at once determine the elasticity and density of aqueous vapour, its weight in a cubic foot of the air, and the degree of dryness, etc., as furnished in the tables in question ; but for the practical man it will be quite sufficient to notice the approach to saturation with moisture in the air, as indicated by the small difference between the temperatures of the dry and wet bulbs—from equality down to about 5° in summer with southerly and westerly winds, but less in winter, and with north-easterly winds in summer. When the difference is beyond 4° or 5° there is very little probability of rain, in spite of the appearance of the sky,—at least, during a long spell of north-easterly winds,—when it often deceives even weather-wise people into the expectation of rain.

With respect to the degree of moisture in the air, an

approximate idea may be formed from the following
table of proportions of humidity for every degree of difference between the two bulbs, at 50° temperature,—
taking equality in the two thermometers as indicating
complete saturation, or 100.

Difference at 50 degrees.	Humidity.
1 degree	93
2 „	86
3 „	80
4 „	74
5 „	68
6	63

It will be as easy to apply these figures in estimating
the probability of rain. For instance, the humidity in
the air on Monday, June 18th, 1866, as well as the day
preceding, closely approximated saturation (96),—the
bulbs being pretty nearly equal in temperature; we had
rain, and it fell heavily in the north-western counties,
the wind being chiefly south-west. On Friday, the
22nd, the humidity diminished to 69, which would infer
a difference of rather over 5° between the two thermometers, and fine weather set in, the wind veering to the
eastward.

No doubt it will be useful to have a few rules for
guidance in prognosticating coming weather by the hygrometer, which is better for the purpose than the barometer, as before stated, due regard being had to the
time of day and the time of year when we make the observations.

1. In summer, when the daily range of temperature
is great, if in the morning the difference between the
air temperature and the dew-point temperature be small,
and the rise of temperature during the day be considerable, it is probable that the difference will increase; and

if the temperature of the dew-point at the same time *decrease*, it is an indication of very fine weather.[1]

2. If, on the contrary, the temperature of both should increase with the day in nearly equal proportion, rain will almost certainly follow, as the temperature of the air falls with the declining sun.[2]

3. In winter, when the daily range of temperature is small, the indication of the weather is shown by the increase or decrease in the temperature of the dew-point, rather than the difference between the temperatures of the air and of the dew-point. If the temperature of the dew-point increases, rain may be expected; if it decreases, fine weather.[3]

4. In showery weather the indications vary rapidly, and if we examine the hygrometer at short intervals we may predict the approach of a storm, especially if we make at the same time observations with the barometer.[4] But the hygrometer, properly interpreted, will be accurate when the barometer may misguide us. Thus, on the 3rd June, 1866, the wind blew strong from S.W. at Corunna and in England, notwithstanding a *rise in the barometer*, which was followed by clouds and copious rainfall, as indicated by the hygrometer; the difference between the bulbs being about 5°.

5. Again: from the 24th to the 29th of June, the wind blew from N.E., with very little (if any) effect on the mean temperature—the thermometer increasing from 61° on the 24th to 75° on the 28th, at Frant, Sussex, as stated by Mr. Allnatt. Now, during the whole interval the hygrometer indicated from 70 to 80, alternately, of moisture, but without rain. On the 29th, however, the temperature fell to 63° from 75°, the hygro-

---

[1] Glaisher, *ubi supra*.   [2] Ib   [3] Ibid.   [4] Ibid.

meter indicated 80 of moisture, and half an inch of rain
fell accordingly, the wind having backed to N., and the
barometer having *risen*—namely, from 29·60° to 29·65°.
This instance also shows that the high amount of humidity (80) was not in itself sufficient to cause rain whilst
the temperature remained high.   So we must watch for
the fall of temperature in prognosticating rain.

6. A high and rising thermometer forbids the prospect of rain, in spite of the large amount of moisture in
the air; but the slightest decrease of temperature settles
the matter forthwith, even with less moisture and a
westerly wind.

7. During the continued sultry weather, such as we
had before the north-easterly winds of July, 1866, the
appearance of big clouds is apt to make us rush to the
conclusion that rain is coming.   It is not the fault of the
clouds if they deceive us, but our present inability to interpret them accurately.   Thus, on the 14th of July, between seven and eight o'clock, the sky became overcast,
and big clouds appeared above, so that a thunderstorm
seemed at hand.   Even the electrometer indicated negative electricity, which is considered a certain sign of
rain.   The only contrary sign was that there was *red* in
the sky, showing that the vapour of the air was not
actually condensed into clouds, but only on the point of
being condensed, and therefore that the weather would
still be fair, according to the Scripture maxim, "When
it is evening, ye say it will be fair, for the sky is red."
And the hygrometer still said there would be no rain as
yet, as, indeed, there was not, for it showed 6° of difference between the bulbs, besides high temperature.

8. The hygrometer will tell us whether wind alone
may be expected from the fall of the barometer.   If the

barometer falls suddenly two or three-tenths, without any material alteration in the temperature, and the hygrometer *does not show much moisture*, a violent gale of wind may be expected.

9. When the hygrometer shows much moisture, with only a trifling fall in the barometer, it merely indicates a passing shower and little wind.

10. When the barometer falls considerably, and the hygrometer shows much moisture, the thermometer remaining stationary or rather inclining to rise, then both violent wind and rain are likely to follow in the course of a few hours, from the southward and westward.

11. If the barometer falls during north-easterly winds, and the difference between the dry and wet bulbs increase or be great (over 5°), expect strong or more wind from the north, before which the barometer will rise, with sky densely overcast and cloudy.

12. A rapid increase in the difference between the bulbs in the morning will indicate a fine day, in spite of all appearances to the contrary.

13. On the other hand, extreme dryness or great difference between the bulbs with a S.E. wind should be suspected. Great rains may follow in a day or two.

14. If in the morning after a fair day the difference remains stationary or decreases, expect rain. This is perhaps the most certain indication of rain.

15. An increase in the difference between the bulbs during rain or snow is favourable at whatever time it occurs.

16. After continuous north-easterly winds, the increase of moisture, as shown by the hygrometer, will clearly show the coming change to the southward and westward, with wet; but then the barometer will have fallen considerably, and the temperature diminished.

17. A slight increase of moisture during north-easterly winds must not be taken as a sign of rain, especially after a warm day, without the indication of the barometer and thermometer.

18. If the difference increases in the evening, after rain, with a south-westerly wind, finer weather may be expected, although the difference be not considerable. The wind will probably back to a drier point—S.S.E. and E.,—but, of course, unsettled weather, as recently occurred, July 28th, 1866, from the backing of the wind.

19. But the most reliable signs are afforded by a progress from day to day towards the moist extreme of the season by the approach towards equality in the bulbs. A retrograde movement towards dryness often takes place during wet or showery weather which the preceding advance towards the moist extreme had prepared us to expect.

Such are the chief rules that may guide the farmer in consulting this invaluable instrument, which, as far as coming rain is concerned, is more reliable than the barometer, because it deals with its immediate cause at every instant—namely, the amount of moisture in the air and the change of temperature.

There are, however, a few variances in the indications of the hygrometer during unsettled weather, connected with *evaporation*, of which more in the sequel.

The instrument should be in universal use, in our changeable climate, not only during haymaking-time and the all-important time of harvest, but also to do away with the many doubts about the weather from the mere *appearances of the skies*, causing the great inconvenience of carrying umbrellas when not at all likely to be required, and their consequent wear-and-tear by carrying, all which might be avoided by investing twelve or fourteen shillings in a good hygrometer, guided by the rules above given.

In speaking thus emphatically of the hygrometer, we are supported not only by practical experience, but by the authority of the most eminent meteorologists, Daniell, Glaisher, and others. The barometer alone cannot suffice to guide us in the prognostication of coming weather; and we repeat that it would be a very great improvement if our meteorological stations were required to give the daily results of the hygrometer, together with the other *data* from which the probabilities of the weather are to be deduced. This is what is wanted; and if people will then attend to the few rules we have given, there will seldom be any mistakes about coming weather, on any of our coasts or anywhere else throughout the country.

Two things must be principally attended to,—the difference between the constituent temperature of the vapour and the temperature of the air, and the variation of the dew-point. In general, the chance of rain or other precipitation of moisture from the atmosphere may be regarded as in the inverse proportion to the difference between the two thermometers, as before observed; but in making this estimate, regard must be had to the time of day at which the observation is made. In settled weather, the dryness of the air increases with the diurnal heat, and diminishes with its decline; for the constituent temperature of the vapour remains nearly stationary; consequently, a less difference at morning or evening is equivalent to a greater in the middle of the day. But, to render the observation most completely prospective, regard must be had at the same time to the *movement of*

the dew-point. As the elasticity of the vapour increases
or declines, so does the probability of the formation and
continuance of rain. An *increasing* difference, therefore, between the temperature of the air and the temperature of the dew-point, accompanied by a fall of the
latter, is a sure prognostic of fine weather; whilst *diminished* heat and a *rising* dew-point infallibly portend
a rainy season. In *winter*, when the range of the thermometer during the day is small, the indication of the
weather must be taken more from the *actual* rise and
fall of the dew-point, than from the difference between
it and the temperature. It must be remembered that a
state of saturation may exist, and precipitation even take
place in the finest weather, and under a cloudless sky;
but this is when the diurnal decline of the temperature
of the air near the surface of the earth falls below an
unfluctuating term of precipitation; and it is probable
that, at some period or other of the twenty-four hours,
this term is always passed. The radiation of the earth,
in the absence of the sun, cools the stratum of air in
contact with it, and a slight precipitation takes place, of
so little density as totally to escape the observation of
the eye. At other times it becomes visible and assumes
the appearance of mist or fog. Under such circumstances, the hygrometer will sometimes exhibit a different kind of action, and the dew-point may be proportionately raised; but in all such cases, the full saturation of the atmospheric temperature must have place,
and consequently, the temperature of the vapour must
be coincident with that of the air. This kind of precipitation, which may often be detected by the hygrometer when it would otherwise escape notice, far from
being indicative of rain, generally occurs in the most

settled weather. It is analogous to the formation of dew, and depends upon the same cause—the radiation of the earth, which can only take place under an unclouded sky. A *sudden change in the dew-point* is generally accompanied by a change of wind ; but the former sometimes precedes the latter by a short interval, and the course of the aerial currents may be anticipated before it affects the direction of the weathercock, or even the passage of smoke.[1]

Even when the indications of the hygrometer are contrary to those of the barometer, reliance may be placed upon them ; but simultaneous observations of the two most usefully correct each other. The rise and fall of the mercurial column is, most probably, primarily dependent upon the state of the upper regions of the atmosphere, with regard to heat and moisture. Local *physical* alterations of its density, thus partially brought about, are *mechanically* adjusted, and the barometer gives us notice of what is going on in inaccessible regions. A *rise* in the dew-point, accompanied by a fall of the barometer, is an infallible indication that *the whole mass of the atmosphere* is becoming imbued with moisture, and a copious precipitation may be looked for. If the mercury falls at the same time that the dew-point is depressed, we may conclude that the expansion which occasions the former has arisen at some distant point, and *wind*, not rain, will be the result. On the other hand, when the air attains the point of precipitation with a *high* barometer, we may infer that it is a transitory and superficial effect, produced by local depression of temperature.

The degree of moisture in the air as shown by the

[1] Daniell, Elem. of Meteorology.

hygrometer, is found by dividing the elastic force of vapour at the temperature of the dew-point by the elastic force at the temperature of the air, when the quotient will express the proportion of moisture actually existing to the quantity which would be required for saturation. Thus, the elasticity of the dew-point at 55° is ·476, and the elasticity of the temperature at 70° is ·770; then ·476÷770=61, the degree of humidity; for, calling the term of saturation 100, as the elasticity of vapour at the temperature of the air is to the elasticity of vapour at the temperature of the dew-point, so is the term of saturation to the actual degree of moisture.

The daily evaporation also may be deduced from the hygrometer. Subtract the elastic force of vapour at the mean dew-point of the day from the elastic force of vapour at the mean temperature of the day, and the remainder will be the day's evaporation. Thus, suppose the mean temperature of the twenty-four hours to be 60° and that of the dew-point 50°, then by the table,—

Elastic force of vapour at 60° = ·560
Elastic force of vapour at 50° = ·400
————
Daily evaporation . . . . . ·160 inch.[1]

Of course these are only approximations.

The air, by its mechanical action, has an influence

---

[1] This is the result of Dalton's *data*, as shown by Dr. Young and applied by Daniell. It happens that the column of mercury equivalent to the elasticity of the vapour expresses accurately enough the mean evaporation in twenty-four hours. Dr. Dalton's experiment gives 45 grains per minute, at the temperature of 212°, from a disk of 3¼ inches. Now 45 × 60 × 24=64,800 grains, or 256·6 cubic inches, which would make a cylinder 30·9 inches in height, on a base 3¼ inches in diameter; and this differs only 1/33 from the height of the mercurial column. We may therefore assume that the mean daily evaporation is equal to the *tabular* number expressing the elasticity of the vapour, sometimes ex-

upon the rate of evaporation. When calm and still, it merely obstructs the process; but when in motion, it increases its effect in direct proportion to its velocity, by removing the vapour as it forms. Dalton fixed the extremes that are likely to occur in ordinary circumstances at 120 and 189 grains per minute, from a vessel of six inches diameter, at a temperature of 212°. Upon these *data* Daniell constructed the following table, which will be found useful for most purposes of the hygrometer, excepting the discovery of the dew-point, which we must either calculate by the " factors " before given, or find by Mr. Glaisher's Tables. As shown, it will enable us to find the existing humidity, and the daily evaporation from the means of the temperature and the dew-point. Besides this, it enables us to discover the force of evaporation at the existing state of the atmosphere. The first column contains the degrees of temperature; the second, the amount of evaporation per minute from a vessel of six inches diameter in *calm weather;* the fourth, the amount in a *moderate breeze;* and the fifth, in a high wind.

To use it, as applied to the hygrometer, proceed as follows. Let it be required to know the force of evaporation at the existing state of the atmosphere. Find the dew-point, as before explained. Subtract the grains opposite that temperature, either in the third, fourth, or fifth column, according to the *state of the wind,* from the grains opposite to the temperature of the air in the same column, and the remainder will be the quantity evaporated in a minute from a vessel of six inches dia-

ceeding it or falling short of it about one-fourth; and we may readily allow for the effect of the moisture of the atmosphere, by deducting the number corresponding to the temperature of deposition.

meter, under the given circumstances. For example, let the dew-point be 55°, the temperature of the air 70°, *with a moderate breeze.* The number opposite to 55° in the fourth column is 2·43; and that opposite to 70° is 3·96; the difference, 1·53 grains, is the evaporation per minute.[1]

TABLE SHOWING THE FORCE OF VAPOUR AND EVAPORATING FORCE ACCORDING TO THE FORCE OF. WIND, BY DANIELL.[2]

Temperature.	Force of Vapour.	Evaporating Force in Grains.		
		Calm.	Mod. Breeze.	High Wind.
18°	·131	0·52	0·67	0·82
19	·135	0·54	0·69	0·85
20	·140	0·56	0·71	0·88
21	·146	0·58	0·73	0·91
22	·152	0·60	0·77	0·94
23	·158	0·62	0·79	0·97
24	·164	0·65	0·82	1·02
25	·170	0·67	0·86	1·05
26	·176	0·70	0·90	1·10
27	·182	0·72	0·93	1·13
28	·188	0·74	0·95	1·17
29	·194	0·77	0·99	1·21
30	·200	0·80	1·03	1·26
31	·208	0·83	1·07	1·30
32	·216	0·86	1·11	1·35
33	·224	0·90	1·14	1·39
34	·232	0.92	1·18	1·45
35	·240	0·95	1·22	1·49
36	·248	0·98	1·26	1·54
37	·256	1·02	1·31	1·60
38	·264	1·05	1·35	1·65
39	·272	1·09	1·40	1·71
40	·280	1·13	1·45	1·78
41	·292	1·18	1·51	1·85
42	·304	1·22	1·57	1·92
43	·316	1·26	1·62	1·99

[1] Similar results may be obtained by Glaisher's Tables, but they differ in some cases.

[2] Elements of Meteorology.

TABLE SHOWING THE FORCE OF VAPOUR AND EVAPORATING FORCE
ACCORDING TO THE FORCE OF WIND—*Continued.*

Temperature.	Force of Vapour.	Evaporating Force in Grains.		
		Calm.	Mod. Breeze.	High Wind.
44°	·328	1·31	1·68	2·06
45	·340	1·36	1·75	2·13
46	·352	1·40	1·80	2·20
47	·364	1·45	1·86	2·28
48	·376	1·50	1·92	2·36
49	·388	1·55	1·99	2.44
50	·400	1·60	2·06	2·51
51	·414	1·66	2·13	2·61
52	·428	1·71	2·20	2·69
53	·444	1·77	2·28	2·78
54	·460	1·83	2·35	2·88
55	·476	1·90	2·43	2·98
56	·492	1·96	2·52	3·08
57	·508	2·03	2·61	3·19
58	·526	2·10	2·70	3·30
59	·543	2·17	2·79	3·41
60	·560	2·24	2·88	3·52
61	·577	2·31	2·98	3·63
62	·594	2·39	3·07	3·76
63	·615	2·46	3·16	3·87
64	·636	2·54	3·27	3·99
65	·657	2·62	3·37	4·12
66	·678	2·70	3·47	4·24
67	·699	2·79	3·59	4·38
68	·722	2·88	3·70	4·53
69	·745	2·98	3·83	4·68
70	·770	3·08	3·96	4·84
71	·796	3·18	4·09	5·00
72	·822	3·29	4·23	5·17
73	·849	3·40	4·37	5·34
74	·877	3·52	4·52	5·53
75	·906	3·65	4·68	5·72
76	·936	3·76	4·83	5·91
77	.966	3·88	4·99	6·10
78	·997	4·00	5·14	6·29
79	1·028	4·16	5·35	6·54
80	1·060	4·28	5·50	6·73
81	1·093	4·41	5·66	6·91
82	1·127	4·56	5·86	7·17
83	1·162	4·68	6·07	7·46
84	1·198	4·80	6·28	7·75
85	1·235	4·92	6·49	8·04

CHAPTER VI.

CLOUD-LAND; ITS COMPOSITION AND COMMOTIONS.

HOWEVER sparkling water may be, it always contains impurities or something else; and however transparent the air, it always contains watery vapour. This fact becomes evident by exposing to the air any cold body, or such substances as are greedy of water,—for instance, potash, chloride of lime, etc. In a word, it is impossible to get dry air without chemical means.

As before stated, the higher the temperature of the atmosphere, the greater the amount of aqueous vapour it contains; and during the rise of temperature, this vapour in it is less and less perceptible. When we lower the temperature of a mass of air charged with aqueous vapour, this vapour, at a certain degree of the thermometer, manifests itself in the form of a cloudiness or nebulosity. Moist air, heated and poured into colder air, determines a nebulosity. Thus, our breath issuing from our mouth in the low temperature of winter or other seasons, produces a little fog visible to everybody. It is the difference of temperature which produces this effect; the same thing happens to the steam issuing from the railway engine at a very high temperature, coming in contact with the cooler air, and forming mimic clouds along the line. Thus, it appears that a difference in temperature of 39·6° between two masses of moist air, which mix, is sufficient to render their aqueous vapour instantly visible; and a difference of 36° in the temperature of *any* masses of air, mixing quietly, is sufficient to exhibit the aqueous vapour they contain in the visible condition.

As soon as the aqueous vapour contained in the atmosphere, or rather, in a part of the atmosphere, becomes visible, it forms a nebulosity and changes state, as established by the experiments of Halley and Saussure. Each aqueous nebulosity, when examined by a magnifying glass, is found to be composed of an infinity of minute vesicles, which any one may see by boiling, in an open vessel, blackened water, or by examining a fog illumined by the sun and projected on a black ground.

When examined through a magnifying glass, a fog is found to be comprised of minute opaque bodies. These small bodies consist of water obeying the law of universal gravitation. The molecules of water group themselves in the form of spherules, just as mercury thrown into a porcelain saucer, or water into a greased vessel. The question is, whether these spherules are full or empty? Halley believed, and it seems more probable, that they are empty, and that the water only serves as an envelope.

On examining the vapour of blackened water heated in an open vessel, Saussure beheld the globules of various sizes traversing the field of his microscope; the smallest passed very rapidly, but a great portion of the others

F

fell down upon the surface of the liquid. According to Saussure, the small vesicles which rise differ so greatly from the others, that it is impossible to doubt the hollowness of the former.

The optical properties of these vesicles confirm this opinion. When full vesicles are exposed to vivid light they scintillate or twinkle. Now, these vesicles never do so. Moreover, it appears that true rainbows have never been observed on clouds composed of vesicular vapour, or not on *drops* of water, rain-drops. But, indeed, Krantzenstein discovered on the surface of these vesicles, rings similar to those of soap-bubbles, and like every German, who is never satisfied unless he goes to the very bottom of every subject, he actually measured the thickness of this infinitesimal envelope. He found it to be 0·06mm., or the six-hundredth part of one four-thousandth of an inch.[1]

[1] The philosophy of soap-bubles is worth investigation, and the following method, suggested by Mr. C. V. Walker, may be acceptable to the curious reader. A simple and very effectual method of preparing a thin film that will endure for several hours, and thus give ample time for studying the order and extent of the several coloured rings or belts is this :—A four- or six-ounce vial is one-third filled with rain or distilled water, and into it is placed a piece of yellow soap, not larger than a pea; the contents of the vial are now brought to the boiling-point, or nearly so, and while, by this means, all the atmospheric air is expelled from within, the vial is closed and removed from the heat, and when cool, is ready for use. For this purpose, give it a short horizontal shake to and fro (the exact mode of which will be learned from a few attempts), so as to form a film across the vial; if cleverly shaken, a single film may be obtained. Such a film is rarely quite parallel to the horizon; it dips slightly, so that the water slowly flows to the lower end; and after a minute or more, coloured bands make their appearance on the upper edge; these travel gently onward and are followed by others; the earlier belts are narrow, the later are much wider. The whole film is soon occupied by the coloured belts, and in due course they all pass onward, and successively disappear at the lower end of the disk.

Both Saussure and Krantzenstein measured the dia-meter of fog-vesicles, and give 0·012mm. as the maxi-mum of this diameter, or the twelve-thousandth part of one four-thousandth of a inch; but this diameter varies in different seasons. In winter, when the air is very moist, the diameter of the vesicles is twice as great as in summer, when the air is dry; but in the same month this diameter also changes. It attains its minimum when the weather is very fine; it increases as soon as there is a threatening of rain; and before rain falls, it is very unequal in the same cloud, which probably con-tains a great number of drops of water, mingled with vesicular vapour.[1] Hence, the measurement of the diameter of these vesicles may be resorted to in pro-gnosticating rain, if the method be not too far above the patient capacity of most observers.

Fog is a certain indication that the air is saturated with moisture; then only can the vapour of water be precipitated incessantly for several hours. It is im-portant to insist on this, for Deluc, and some other meteorologists, who have employed imperfect hygro-meters, have insisted that the air is often very dry in

The respective colours pass into each other through the intermediate tints. After the coloured belts there arrives a broad belt of white, which is succeeded by a black belt, which terminates the phenomena, but the transition from the white to the black is sudden, and is dis-tinctly defined by a straight line. Words are inadequate to express the beauty of these tints when viewed by reflected light, or looked at; nor is language able to enumerate the various hues. When viewed by transmitted light, or looked through (and this requires a little manage-ment), the complementary colours are seen. I have had a film of this kind in existence for ten or a dozen hours, at the end of which period, all the coloured belts and the white had passed away, and nothing remained but a black disk, which, by transmitted light, was white. (Note by Mr. Walker, Kaemst's 'Meteorology.')

[1] Kaemst.

regions where fogs are forming. The experiments of Saussure, however, prove the contrary, and Kaemst confirms this fact from experiments on the Alps and in different parts of Germany. An hygrometer suspended before a window in the centre of a city, undoubtedly cannot indicate the degree of saturation during the times of fog, and this happens because the instrument is warmed by the walls of the building; but even this anomaly disappears when the fog remains for several hours.

The circumstances in which fog forms, are often very different from those which accompany dew. When dew is deposited, the soil is always colder than the air. When fog occurs, the contrary is the case; then the moist soil is warmer than the air, and the vapours that ascend become visible, like those which rise from boiling water, or like the vapour of expired air, before mentioned, which, in winter and other times, condenses the moment it escapes from the mouth. So, in autumn, fogs frequently appear above rivers, the water of which is much warmer than the air before sunrise.[1]

But fogs demonstrate that the atmosphere must be *moist*, already saturated, otherwise it would absorb the vapour of fog and render it invisible. The ancients made an observation, the truth of which may be verified even in our own times. When the crater of the volcano Stromboli is covered with cloud, the inhabitants of the Lipari Isles know that it will soon rain; and the reason is, because the air, already saturated with the vapour of water, cannot dissolve that which escapes from the crater. In like manner, the inhabitants of Halle prognosticate rain when the vapour of the salt-springs, near

[1] Kaemst.

them, covers the city, simply because the air, being saturated with vapour, cannot absorb them.

In countries where the soil is moist and hot, and the
air moist and cold, thick and frequent fogs must be
expected. This is the case in England, the coasts of
which are washed by a sea at an elevated temperature.
The same is notoriously the case in the Polar seas of
Newfoundland, where the Gulf Stream, which comes
from the south, has a higher temperature than the air.

A word about London Fogs will here be appropriate,
although of late years they have not maintained their
character of British " institutions," more or less connected with suicides in gloomy November. Often has
it been necessary to light the. gas in the middle of the
day in the streets and houses. On the 24th of February,
1832, the fog was so thick, that at midday, people in
the street could not see distinctly; and in the evening,
the town having been illuminated to celebrate the birth-
day of Queen Adelaide, boys went about with torches,
saying they were looking for the illumination.

Something more than the "vesicular vapour" of
which we have been speaking has to do with London
fog. London fog is a monstrous cross betwixt vapour
and London smoke; and no doubt it is to the extensive
abatement of the smoke nuisance that we are indebted
for the mitigation, to a great extent, of London fog in recent years. A little explanation will enable us thus to
trace the old evil to its cause. If incandescent or burning charcoal dust be allowed to cool *in vacuo*, and is
then immediately placed in a globe containing any gas,
this gas is absorbed, especially if it be charged with vapour of water. The carbon increases even sensibly in
weight, so that 100 lbs. of incandescent charcoal, when

exposed to the open air, will weigh, at the end of several days, 210 or 214 lbs. This fact is well known in manufactories of gunpowder.

In escaping, therefore, from the chimneys, the particles of carbon in London smoke must absorb air, and increase in weight. The wind, however, will carry them some distance before they fall to the earth; but if the air is *moist and calm,* as is the case in fog, the specific weight of the particles rapidly increases; they mix with the fog, and are diffused with it into the neighbourhood. Such is, or was, the *rationale* of London fog, which, thanks to the Act of Parliament against the smoke nuisance, is, we trust, a ghost laid at last for ever; at least never again to re-appear in its old solid proportions.

People sometimes talk of *dry* fogs. This is a contradiction in terms, an utter impossibility, unless it refers merely to smoke or clouds of dust. A fog is necessarily moist, even without being positively a *Scotch mist,* which is about the best representative we can think of for *saturation with moisture.* The composition of these so-called dry fogs is not moisture. In long-continued hot, dry weather, the air is generally dense and filled with various mineral and earthy particles, together with seeds and animalcules, which, from *want of moisture* to increase their gravity, are enabled to float about in the atmosphere, and are supposed by some writers to generate epidemical disorders. Extensive dry fogs prevailed over Europe in the year 1783, and numerous instances are given in historical and chronological works of similar phenomena.

Heavy rain, especially when accompanied with thunderstorms, clears the air and terminates this dry fog or rather mistiness; and this rain explains its origin. According to Brandes and Zimmermann, it contains a great variety of substances—iron, manganese, nickel, salts of magnesia, soda, lime, and other ingredients. There are many instances of hailstones which contain metallic particles; and Ehrenberg discovered in the dust that fell at the distance of 380 miles from the coast of Africa, the remains of eighteen species of "siliceous-shelled polygastric animalcules."

Moreover, the misty appearance of the atmosphere during very hot dry weather, will be increased by the mixing of airs of different densities, owing to the great variations of the temperature of the surface of the arid soil, owing to the cloudless days and nights. Before sunrise the ground is rendered very cold by radiation or parting with its heat during the night, and the air above it, to a certain height, has a much reduced temperature. When the sun rises, as the earth becomes heated, the air above it expands upwards, and by mingling with the air of a greater density, the result is, that the different refracting powers of the two uniting gases prevent the direct passage of the sun's rays, and cause the opaque appearance called dry fog or mistiness. This is similar to what occurs in a mixture of alcohol or strong whisky and water before they are completely united, the "cloudiness" of which is apt to make us suspect some impurity in the ingredients.

Fog is one of the greatest annoyances that the arctic whalers have to encounter. It frequently prevails during the greater part of the month of July, and sometimes, at considerable intervals, in June and August. Its density is often such, that it circumscribes the prospect of an area of some acres; not being pervious to the sight at the distance of a hundred yards. It frequently lies so low that the brightness of the sun is scarcely at all intercepted; in such cases, substances warmed by the sun's rays, give to the air immediately above them increased capacity for moisture, by which evaporation goes briskly on during the densest fogs! In Newfoundland, on occasions when the sun's rays penetrate the mist and heat the surface of the rocks, fish is frequently dried during the thickest fogs.

These Arctic fogs are more frequent and more dense at the borders of the ice near the coast of Spitzbergen. They occur principally when the mercury in the thermometer is near the freezing-point; but they are by no means uncommon with the temperature at 40° or 45°. They are most general with south-westerly, southerly, and south-easterly winds. They seldom occur with high winds, yet, in one or two instances, Scoresby observed them very thick even in storms. Rain generally disperses them.

In the Arctic regions, fogs, by increasing the apparent distances of objects, appear sometimes to magnify men into giants, hummocks of ice into mountains, and common pieces of drift-ice into heavy floes or icebergs. They are an especial annoyance to the whale-fisher, and greatly perplex the navigator by preventing him from obtaining observations for the correction of his latitude and longitude, so that he often sails in complete uncertainty. Arctic fogs are more common near the ice than in the vicinity of the land, more frequent in open than in close seasons, and more intense and more common in the southern fishing-stations than in the most northern.*

Analogous to fog or mist in their nature, the clouds

* Scoresby, ' The Arctic Regions.'

float in the atmosphere, which is their ocean. Before considering them a few observations may be useful.

Above the terminal surface of the vapour-atmosphere, before described, there exists, in the upper regions, a certain quantity of vapour in the invisible state, as proved by the experiments of Saussure, Kaemst, and others. Does this vapour rise directly from the surface of the earth with the ascending currents previously mentioned, or is it simply an emanation from the terminal nebulosity of the vapour-atmosphere? It is difficult to come to a decision on the subject; perhaps both causes contribute to the result in question; for the present purpose, however, it may be sufficient to establish the existence, above what is conventionally called the atmosphere of aqueous vapour, of a second mass of vapour, less dense than the former,—at any rate, less thick, generally more transparent, and like the former also terminated above by a horizontal surface.

It is in these two spherical envelopes of vapour, one above the other, the lower of which touches the surface of the earth, that occur all the phenomena which concur in the formation of rain, the commotions of " the elements."

According to M. Rozet, the temperature of the first of these envelopes never gets lower than 32°, at all events in our regions, excepting in winter.

In the second, on the contrary, the temperature never gets above that point excepting in the vicinity of its lower surface, and also probably during the hot season. At its upper surface, the thermometer is often as low as 5° in summer, and the vapour in it is in the frozen condition. Such is the grand laboratory of Nature, where she prepares sunshine and showers and all their accompanying phenomena.

In all the books treating of meteorology, we are told that a cloud is nothing but a fog or mist, the observer being situated *outside* of it; in other words, that any aqueous nebulosity is a fog or mist to the observer plunged in it, whereas it is a cloud to the observer outside of it; so that, according to this definition, there is no difference between a cloud and a fog or mist. With M. Rozet, we believe this to be an error. The clouds may be really very little more than flimsy "nothings" in spite of their very grand and imposing dimensions on certain occasions, but they nevertheless lay claim to a well-defined individuality.

A *fog or mist* is a region of the atmosphere, whose temperature is so much lower than that of the ambient mass—including the soil—that the greater part of the vapour in it, or that which comes to it, passes from the invisible into the visible state.

A *cloud* is the grouping of visible aqueous vapour assuming a definite form; permanent, at least for a certain time, endowed with special properties which constitute an individual; lastly, resisting, with certain energy, the external forces which tend to destroy it.

Peltier accurately described a cloud, and its composition is as follows:—It consists of opaque globules grouped or arranged in minute flocks or flakes, having their limits and spheres of action, like the globules themselves. These minute flakes, in grouping, form larger flakes; the latter group and form *mamelons*, or breasts; a certain number of mamelons, by uniting, form a *nuelle*, or small cloud; and these *nuelles*, uniting, constitute the definite clouds, as we see them in all their fantastic variety.

The people of all countries had all along distinguished

the various kinds of clouds, and given them characteristic names; but Luke Howard thought proper to give the subject a scientific form, classified the clouds, and re-christened them, giving them a set of barbarous names, simple and compound, viz. Cirrus, Cumulus, Nimbus, Cirro-cumulus, etc., which Mr. Glaisher has trans-mogrified into Ci, Cu, Ni, Ci-cu, etc., which sound ex-ceedingly like the dialect of certain North American Indians, or the cackling of the African Bushmen. For want of a better (we suppose) Luke Howard's cloud-nomenclature has been adopted by the meteorologists of all Europe, and so it is absolutely necessary to retain it in speaking or writing on the subject.

Howard's list, however, may be abridged into his three principal forms,—the *cirrus*, the *cumulus*, and the *stratus*, which become modified into compound shapes, which he terms *cirro-cumulus*, *cirro-stratus*, *cirro-cumulo-stratus* or *nimbus*.

The *cirrus* is a cloud resembling a lock of hair, or a feather, with parallel, flexuous, or diverging fibres, un-limited in the direction of their increase. It is always the least dense, and generally the most elevated modi-fication of clouds, sometimes covering the whole face of the sky with a thin transparent haze, and at other times forming itself into distinct collections of parallel threads, wavy, or diverging fibres. It is supposed to consist of icy spiculæ, on account of its great elevation, which is above the altitude of congelation.

The *cumulus* clouds are those huge clouds with rounded shape, terminated below by a horizontal surface, appear-ing in fine summer weather especially, and forming masses more or less considerable in the various regions of the sky.

The *stratus* is an extended, continuous, horizontal sheet of misty particles,—the sort which may be observed towards sunset, and destined to disappear at sunrise, or to change into *cumulus*. As Becquerel observes, the cumulus may be considered as a day-cloud, and the stratus as a night-cloud.

In the *nimbus* all the forms are so mixed that we cannot recognize any of them. It may be termed a shapeless cloud. It is a thick fog.

The other forms of clouds explain themselves by their compound names, having reference to the elements of their composition.

One would like to observe a cloud in its process of formation; but the thing seems to be impossible. De Saussure once thought he had a good opportunity for doing so, but he was disappointed.

" Stopped by a rainy wind on the top or the slope of a mountain," writes Saussure, " I endeavoured to catch a glimpse of the formation of the clouds, which I could see coming into existence almost at every instant, over the forests or the meadows situated beneath me. No fog covered the surface; the air which surrounded the woods and meadows was clear and transparent; and yet, suddenly, sometimes on one hand, sometimes on the other, clouds made their appearance without my being able to detect the beginning of their formation. In a part from which I had just moved my eyes—where two seconds before there was no cloud—I suddenly beheld a tolerably big one."

When rain is at hand, the clouds increase in thickness without a rise in the thermometer, by the addition of a new quantity of vesicular vapour.

The temperature of the interior of a cloud formed of

vesicular vapour is always lower than that of the surrounding air; but it is otherwise with storm-clouds, whose arrival lowers the temperature considerably.

When a layer of cumulus, formed of isolated clouds, persists, the latter hurry up together and come into contact, to form a mass more or less continuous. But when the clouds touch they never become confounded; the contact is similar to that of elastic bodies pressed one against the other. M. Rozet frequently had the opportunity of verifying this curious fact. "Placed above them," he says, "I have often observed, with a good telescope, the cumulus clouds when they touch each other, and I never remarked the slightest electric discharge between them. This fact, together with the species of repulsion between their lateral surfaces—pressed one against the other—proves that they are all in the same electrical state. Being above a layer of cumulus, in a clear and cloudless atmosphere, I saw them persisting for several days successively, without yielding the least drop of rain,—hiding the sun from all the country below me. When an opening occurred anywhere in the mass, the sun's rays passed through, and I saw beneath me the entire landscape perfectly illumined."

Before rain, the cirrus clouds crowd together and form cirro-cumulus. This gives rise to what is called a "dappled sky." Hence the French rural proverb:— *Temps pommelé, femme fardée, ne sont pas de longue durée,* that is, "a dappled sky and a painted woman don't last long."

The cirrus clouds repose on a regular spherical vault, like the cumulus. This vault must terminate a second mass of vapour, comprised between the upper surface of that which produces the cumulus clouds, the region where aqueous vapour freezes.

Thus there are two principal regions in the atmosphere for the formation of clouds :—

1. That of the cumulus clouds, in which region the vapour that rises from the surface of the earth passes into the vesicular state, and may attain an altitude of about 12,000 feet.

2. That of the cirrus clouds, in which region vapour freezes, and the height of which is, in summer, certainly not less than 15,000 feet.

Hence, there are really only *two* kinds of clouds, the cumulus, formed of vesicular vapour, and the cirrus, formed of frozen vapour. All the other kinds of clouds distinguished by Howard, are merely modifications of these, as M. Rozet and everybody have found to be the case in a multitude of observations.

The cirrus, also, must be in the same electric state, for, situated above the layer of cirrus, and observing with a good telescope, M. Rozet never saw the least discharge between cirrus clouds when they touched each other, and even when they mingled, which is often the case in the filamentous or thready condition of these clouds.

In the state of equilibrium, the upper surface of the cumulus clouds never reaches the lower surface of the cirrus clouds. When both of them exist together in the atmosphere, there is always an open space, a considerable interval (not less than 3000 feet) between the two layers formed by each of these kinds of clouds. If, throughout a layer, the clouds are in the same electrical state, each layer has a different electrical condition. M. Rozet often witnessed strong electrical discharges between the cirrus and the cumulus which touched each other in a tempest. Sometimes the discharges are very weak, although these clouds come in contact. He frequently failed to perceive any trace of a discharge, although observing with an excellent telescope. Thus, then, the quantity of electricity in the clouds must vary remarkably, according to certain circumstances, respecting which we have no positive data.

We now come to the consideration of those commotions of the atmosphere, which are manifested in storms and tempests of rain.

There can be no doubt that electricity plays an active part in all storms. Saussure never witnessed any in the mountains without the approach and conflict of two or more clouds. A storm is therefore a battle of the clouds, and they seem to "go at it" somewhat after the well-established human fashion, in all times, from the beginning. As long as a single cloud, or a single kind of clouds, remain in the skies, however dense, dark, and lowering they may be, there is no thunder; but when two layers form one above the other, or one of them mounts up from the plains or the valleys, to get at those which occupy the heights, then they set to, and exhibit all the grand results of "heaven's artillery," thunder and lightning, hail and rain.

The Commandant Rozet's experience with these celestrial belligerents is precisely similar. He witnessed many of their pitched battles on the Alps and the Pyrenees, and describes them with a sort of professional feeling, as becomes a French warrior. As long as there existed above or below him a single species of clouds, either cumulus or cirrus, forming a layer more or less dense, whatever might be their thickness and compact-

ness, there was neither tempest nor rain, although the
layer, chiefly cumulus, persisted, for several days successively, in occupation of its ground—if the expression
be allowed—rising and falling day by day with the sun.
Again, when there were two different layers, cumulus
and cirrus, as long as they kept at a certain distance
from each other, fine weather still continued. Sometimes, but rarely, these two layers would disappear
under the influence of the sun's rays, and then, of
course, the weather remained fine; but when they persisted in presence, the clouds of the upper layer were
never long in descending, whilst those of the lower layer
mounted in columns of attack, more or less irregular,
and their encounter began with the formation of the
nimbus, which would appear to be the skirmishers of the
clouds, generally increasing very rapidly, and often,
especially in summer, accompanied by lightning and
thunder, rain falling very soon after.

Long ago, Hutton remarked that when two masses
of saturated air, but of unequal temperatures, come
together, there is a precipitation of aqueous vapour. If
the two masses of air are not in a state of saturation,
they nevertheless become more moist, and if the temperatures are very different, there will be a precipitation
of vapour, even although the two masses of air be not
saturated. Now, M. Rozet was often on the summits
of the Alps, in the very spot where the cumulus and
the cirrus came together, that is, in the nimbus whilst
forming, and he found that the temperature suddenly
fell 55° and even 59°, a proof of the low temperature of
the cirrus, which came from a much higher region.

Monck Mason, in his aeronautic excursions, observed, that when there is rain from a sky completely

covered with clouds, there is always a similar range of clouds situated above at a certain height; and that, on the contrary, when it does not rain, although the sky presents below the same appearance, the space situated immediately above presents, as a dominant character, a great extent of clear sky, with a sun unobscured by a single cloud. This explains why a similar state of things frequently exists, a very cloudy, overcast sky without a drop of rain.

According to Peltier, a thunderstorm is composed of two ranks of clouds, one "resinous" or "negative" below, and the other "vitreous" or "positive" above. It is almost always between these two layers of clouds that electric discharges take place. The discharges between negative clouds and the earth are much more rare than is generally believed.

According to Kaemst, a thunderstorm is formed several hours before it bursts forth. In the morning the sky is perfectly clear; towards midday we see isolated cirrus, giving the sky a whitish aspect, the sun being pale and of a sickly hue, with "parhelia" or mock suns and haloes round about him; later, the cumulus clouds make their appearance, and, in extending, confound themselves with the upper layer. Shortly before the storm bursts, a third layer is seen, which appears especially in mountainous countries.

The formation of thunderstorms and rain are two phenomena, which are always intimately connected together, and they result from the encounter of the cirrus and the cumulus, in other words, of frozen vapour and vesicular vapour. It does not appear possible that one of these vapours alone can produce true rain.

The great degree of cold produced by the meeting of

these clouds explains the lowering of the temperature during rain, which has been often prejudicial to the crops in rainy seasons. Frequently in the Alps and the Pyrenees, during a hot summer's day, M. Rozet was frozen in his observations at the sudden arrival of a storm cloud, and warmed again by the sun ten minutes afterwards.

This low temperature of the nimbus determines currents of cold air from the spot which they occupy, towards the surrounding lower regions, which are warmer.

It sometimes happens, especially in autumn and in winter, that after rain, light blackish clouds separated from each other, circulate between the cumulus and the surface of the ground, producing rime or drizzling rain. If sunshine and fine weather are to follow, these clouds soon disappear; but should the showers be likely to continue, these clouds become more numerous, thicker, and finish with raining themselves; their vapour is then probably precipitated by the water which falls from the upper nimbus.

The mercury in the barometer begins to fall as soon as the cirrus and cumulus appear together in the atmosphere; but it is at the moment of the formation of the nimbus that the fall is greatest. If the rain only lasts a few hours, the barometer remains stationary all the time the rain falls, and rises immediately afterwards, whilst, at the same time, the nimbus rises in packs.

From all that we have said a few trustworthy weather-signs may be deduced.

1. When fogs and mists ascend, rain cannot be far off; for, in effect, they are a fresh quantity of vesicular vapour added to that already contained by the atmo-

sphere, and must materially augment the volume of the clouds.

2. The sun setting in clouds, after fine weather, is a sign of coming rain; for this fact attests the formation of clouds in the region of the west above the ocean, whence our storms arise.

3. Circles, haloes, etc., round the heavenly bodies are prognostics of coming rain, although there be no clouds in the skies, for they attest the saturation of the upper regions of the air with vapour; and only let a south-west wind bring up the cumulus, and we shall soon see the skies overcast with rain-clouds forthwith.

4. The "dappled sky," before mentioned, is a certain sign of rain, because it consists of combined cirrus and cumulus, or *cirro-cumulus*; in other words, a mixture of icy vapour and vesicular vapour. Betwixt the dapples the sky will be grey, or rather whitish, very rarely blue. Here we have the two elements of rain, the icy vapour uniting with the vesicular, and so a downpour is certain.

5. A pale sun in the morning announces rain, especially when cumulus clouds appear in the south-west; for his rays are weakened by having to pass through a thick whitish layer of frozen vapour, in the midst of which will soon appear cirrus clouds, whose union, during the day, with cumulus, the numbers of which increase as the sun ascends, produces bad weather.

6. Hoar-frosts in spring and autumn are generally followed by rain in twenty-four hours; for they are produced, as before explained, by a rapid evaporation of the dew, which deprives itself of heat to congeal on the plants, etc., a portion of the water deposited on them; the consequence is, that a large quantity of vapour

is suddenly carried into the air, whence it must soon fall
down again in rain.

7. Great clearness of vision in the air is a sign of
rain; for the air is then saturated with vapour, at least
in the region of the atmosphere where vesicular vapour
can exist, and such air is more transparent than dry
air. The cause is probably connected with the fact
that saturated air precipitates by wetting them, the
particles of matter, not gaseous, which float abundantly
in the air up to a certain height; for we cannot admit
that moist air absorbs less light than dry air, since,
near the horizon, the light of the sun is weak enough
to allow us to look at it with the naked eye. If the
air be thus mechanically cleared of the obstacles to distinct vision by its superabundant moisture, this condition must be favourable to rain when it approaches saturation.

8. When the sky is veiled with a layer of cumulus, if
we perceive everywhere the azure blue in the intervals
between the clouds, there will be no rain; but as soon
as, and whenever, instead of the blue we perceive a grey
tint, more or less decided, or filaments of cirrus, we may
predict rain.

9. When the sky is completely hidden by a layer of
cumulus clouds so close together that no opening between them can be seen, we need not apprehend rain as
long as that layer does not descend, and as long as the
lower surface preserves its regularity, and provided it
remains composed of parts more or less rounded; but
when the clouds descend—which sometimes happens
suddenly—the level of the lower surface is immediately
destroyed, various movements are seen in it, nimbus
clouds are formed, and bad weather will soon set in.

10. When, through a layer of nimbus raining or snowing, we can perceive through the breaks in it the blue of the sky, or an unequal white tint sprinkled with blue, we may predict the early cessation of bad weather. Then the nimbus rises in packs, the intensity of the rain diminishes, and finally ceases. Cumulus clouds then form here and there near the surface of the ground, especially in mountainous countries.

Great as are the ravages of storms in our latitudes, their violence is not to be compared to those of the hurricane and other tempests of the torrid zone, or the furious blasts of the arctic regions. All are familiar with accounts of the equatorial tempests, but few may know that they are equalled in intensity by the storms of the high northern regions. Here the wind rages so vehemently at times, that the houses quiver and crack, the tents and lighter boats fly up in the air, and the sea-water scatters about the land like snow-dust,—nay, the Greenlanders say that the storm rends off stones a couple of pounds' weight, and mounts them in the air. In summer, whirlwinds spring up that draw up the waters out of the sea, and turn a boat round several times.

On the other hand, when the countries of temperate climates suffer under tempests in frequent succession, Polar regions enjoy comparative tranquillity. After the autumn gales have passed, a series of calm weather, attended by severe frosts, frequently succeeds. So striking, indeed, is the stillness of the northern winter, that there is truth in Dr. Guthrie's observation, that nature seems to have studied perfect equality in the distribution of her favours, as it is only parts of the earth which enjoy the kindly influences of the sun that suffer by the effects of its superior heat; so that, if the atmosphere of the north is not so genial as that of the south, at least it remains perfectly quiet and serene, without threatening destruction to man and the product of his industry, as in what are called happier climates.

CHAPTER VII.

THE CURIOSITIES OF RAINFALL IN ENGLAND AND IN EUROPE IN GENERAL.[1]

SEVERAL cycles or periods of maximum and minimum rainfall have been devised by ingenious investigators; but, after a very elaborate examination of the subject, Mr. Symons has been unable to attest the accuracy or sufficiency of any of them. He has, however, discovered the important fact that the wettest years have run in *twelve*-year periods ;—1860, 1848, and 1836 had a mean of 34·27 inches, which is *higher* than any other individual year except 1852.

The *dry* years, on the other hand, seem to run in *ten*-year periods ;—1864, 1854, 1844, and 1834 had a mean of 23·18 inches, which is *lower* than any other individual year, except 1858.

But then there have been two consecutive " wettest " years (1848 and 1849), and even three " wettest " years (1828, 1829, 1830); also two and three " driest " consecutive years (1854 and 1855 ; 1854, 1855, and 1856).

At the present time, when so much interest is felt on

---

[1] A portion of this chapter appeared in the ' Mark Lane Express.'

all hands respecting our excessive drainage, diminished water supply, and the condition of our rivers, the following results, deduced by Mr. Symons from an examination of the rainfall of fifty years (from 1815 to 1864), in ten widely separated stations, may at least be interesting if they do not lead to any definite inferences as to the periodicity of maximum rainfall. The stations are in the following counties, Devon, Kent, Middlesex, Surrey, Essex, Lincoln, Lancashire, York, Edinburgh, and Argyle.

In the first ten years (1815 to 1824) *seven* were above the average of fifty years; in the next ten (1825 to 1834) *six;* in the next (1835 to 1844) *five;* in the next (1845 to 1854) *four;* in the last (1855 to 1864) *three.* In the first twenty-five years *sixteen* were *above* the average, and in the next sixteen were *below* it.

This fact seems to infer something like periodicity, involving apparently that compensation or tendency to equilibrium which characterizes all nature; but there may be another cause for it, which will be mentioned in the sequel.

From the means of these fifty years, classed in periods of five years, Mr. Symons reasons as follows:—"From these tables we find that, when, as in this case, local irregularities are neutralized by the combination of observations from a large tract of country, rainfall records evince a regularity not before expected, the main and marked feature being the drought in the years 1854 to 1858. *Omit* these five years, and then the records run in *five*-year means, without a single departure of an inch from the average. But it will not do thus to *omit* them; they were exceptional, but are part and parcel of the whole, and must by no means be separated; the whole must be carefully examined. For several reasons it seems better to take the ten-year means; and from them we find that the annual fall in each ten years, from 1815 to 1854, was *nearly equal*, and always greater than in the last ten years (1855 to 1864); and, moreover, that the ten years 1845 to 1854 had a rainfall (28·61) nearly identical with the mean (28·60) of the preceding thirty years (1815 to 1844). Hence, it is evident that at any stations where observations have been made continuously from 1845 to 1864, we may take the ten years 1845 to 1854 as representing the forty years 1815 to 1854, and the difference between the first and last ten years as representing approximately the decrease of rainfall at that place."

According to Mr. Symons, the leading features at present seem to be—(1.) A decrease averaging four per cent. over the whole British Isles, but unequally distributed, the decrease being exchanged for an increase in parts of Ireland and the south of Scotland. (2.) In England, although the amount of decrease varies up to 18 per cent., it never falls below an excess (if the expression may be allowed) of 2 per cent. (3.) Although at first the figures seemed very discordant, yet, on drawing the lines on a map some order seems to become evident—namely, that the maximum deficiency has existed along a line running S.W. to N.E. from Cornwall to the Wash. Proceeding north-westward, the deficiency becomes less, until the parallel line running through the centre of Ireland, and passing into the North Sea at Edinburgh, marks a district in which no deficiency has existed, but, on the contrary, an excess of nearly 10 per cent. This is, we believe, a striking fact, and we shall revert to it in the sequel.

The next districts follow nearer to each other, and seem to involve the eventual adoption of west to east, instead of south-west to north-east as at first apparent. Possibly this is not the case; but the difference is rather due to errors of observation at the lighthouses, whence most of the values which are assigned for Scotland are derived; or it may arise from the modifying influence of Ireland not being felt in those higher latitudes. "But," said Mr. Symons, addressing the British Association for the Advancement of Science, "I might further point out that the deficiency seems in some degree connected with the large drainage operations in the middle and eastern counties of England; but until, either by my own efforts, or the assistance of the Association, the observations are rendered more complete, it is not safe to attempt to determine the causes of the recent fluctuations."

We incline to believe that the last suggestion is the probable cause of the deficiency; for certainly in no period has man's activity been so great as during the last ten years in altering the face of the country, by railway cuttings, by clearing out spaces, by agricultural and other drainage, by the operation of huge manufactories consuming prodigious quantities of water, and turning still more to waste. But, indeed, these causes have been in operation during the entire period of the last twenty-five years, wherein, as before shown, the rainfall has been below the average, excepting in those localities which have been exempt from these effects of advancing "civilization." And here we must again turn to the line of maximum, of excessive rainfall, so curiously discovered by Mr. Symons,—a most important fact, as we take it, in meteorological investigation.

This line of maximum and increase of rainfall begins at Valentia, in Ireland, in the S.W., proceeding N.W. between Dublin and Belfast, crossing the lake districts of England, and through the Border Land to the German Ocean. Now, throughout this belt of copious rainfall all the conditions of vegetation, tall trees, and elevated points exist, and man has done little or nothing to interfere with Nature. We believe it will be found to be the same everywhere else; the more natural the locality the more will Nature favour it with the fruitful rain, which is the milk and honey of the land.

But if we cannot undo the past, is there no remedy for the future? Can nothing be tried to restore our waters by increased rainfall in addition to checking waste? Let us consider the proposition. Rain clouds are attracted to certain localities more than others; and probably Nature's conductors are the points of the leaves of all vegetation, particularly trees; and hence, to the cutting down of trees may be traced their ultimate sterility—such as the present sterility of the once most fertile, but now deserted and desolate regions of Syria, Barbary, and Chaldea. The Euphrates often menaced ancient Babylon with inundation; but at present, thanks to the clearance of the woods from the mountains of Armenia, the river is modestly confined to its banks. The ancient river Scamander, which was navigable at the commencement of the Christian era, has completely disappeared with the cedars of Mount Ida, where it took its rise. If trees do not prevent the drought of countries, their roots open the soil for the percolation of water, and they oppose the scattering of the sands of the seashore upon the plains—another cause of inertility.

Such are the results of depriving a country of its
trees, and the question is whether we cannot take a hint
from Nature, so as to make amends for having interfered with her provisions. As suggested long ago by
Sir Richard Phillips, should we not be able to prevent
these disastrous droughts with which we have been of
late so often visited, by erecting *metallic conductors* over
the country, which might even be more effective than
trees, besides vastly diminishing the chance of suffering
from the thunderstorms? Placed on elevated surfaces,
a sufficient number of such conductors would arrest the
clouds, and produce sufficient rain to sustain vegetation,
and refill the almost exhausted rivers in the most barren
regions.

On the other hand, it may not be useless to suggest
the propriety of reverting to our old custom of " culti-
vating " the oak. British oak will always be in demand,
and it is now by no means superabundant. Then, not
only on its own profitable account, but with the direct
object of increasing our rainfall where deficient, let it be
planted on the hilltops, and every locality unfit for other
cultivation. Some trees have a natural aptitude, apparently owing to their moisture, to attract the thundercloud, and the oak is one of them. Many an owner of
wide lands should take this hint, and become a bene-
factor to himself as well as the nation, which is the best
way of making both ends meet here below.

Artesian wells would, of course, be another means to
supply the deficiency of water, and indeed the only
means of obtaining a supply sufficiently large to meet
the requirements of effective irrigation, if this is to be
done artificially. There is no comparison between the
quantity of water that falls in an hour's steady rain,

and that which could be applied by any artificial means at present in our power. A fall of rain measuring only one-tenth of an inch seems a small matter, and yet to equal it artificially would require about ten tons per acre, or 2262 gallons of water! But take another illustration: Half an inch of rain is a very moderate amount for a thunder-shower. Well, how many cans of water would be required to equal half an inch on a garden of 22 yards square (a tenth of an acre)? If your can holds 4 gallons, you will have to fill it no less than 282 times, and pour out no less than *five tons* of water! Every gardener knows that watering is useless in drought unless it be copious and drenching.

In its normal condition the rainfall of our island has been very varied in its amount in different localities, depending upon the peculiar features of the country, as before stated. In the lake districts, for instance, the yearly fall is sometimes as much as 150 inches or over; in the extreme south of England it averages from 30 to 31 inches; in the neighbourhood of London, 25 inches; about 24 inches in the midland counties, and considerably less in some localities; 30 inches on the west coast, which, in the hilly districts of Cornwall and Devonshire, is sometimes increased to 40 inches, and about 26 inches along the east coast. As a general rule, the rainfall is greater on the west than on the east coast, and on mountains than on plains, precisely for the reason we have given, namely, the power of the innumerable points of trees and minerals to disturb the electricity of the clouds, and make them fall in rain.

Were it not for the late disturbance in the averages throughout the country, we might at any time foresee approximately how much more rain will fall during the

year by deducting the quantity fallen, as recorded by
the rain-gauge, from the average of the locality. In
some seasons, when the hay-harvest is early, the removal
of the immense exhaling surface of grass has an influence
on the weather, particularly if a few bright days immediately succeed, as a check is then given to the after-
math; and during this interval between the hay and
corn-harvest, we may have fine weather for filling and
ripening the grain. The produce, in such cases, is generally abundant.

There is an old proverb, " Drought never did hurt to
England," which has puzzled our moderns; but the
axiom is easily explained if we consider that before England was extensively *drained* as it is now, a drought was
a benefit to the saturated land. Such is the meaning of
the proverb.

England does not suffer from a *wet* harvest above once
in *six or seven* years; and when even this happens, it
arises from some *general cause*, not from local influence;
and then our Continental neighbours experience the
calamity, partially, if not equally with ourselves. Thus,
the last wet harvest (1866) may be traced back to its
likeness in 1799, when the spring commenced in April,
with frequent storms of snow; May and June were remarkably cool and cloudy; July and August, even cold
and wet; the grain turned colour, the weather seemed
labouring to clear up for the first eleven days in September, when the rain again commenced, and so continued
till the winter frosts set in, which were very severe; for
on the 30th and 31st of December, the thermometer was
as low as 10°. Owing to defective drainage, the following seasons were most untoward; the grain was very deficient in quantity and worse in quality; so that if it had

not been for the exertions of Government in promoting the import of grain, rice, etc., the country must have experienced all the horrors of famine. After unfavourable seasons it would be prudent to sow *two-years' old* wheat, rather than such as is imperfect and liable to be unproductive.[1]

According to a table before us, on an average of 46 years, it appears that some rain falls on 136 days out of the year, and that the average is divided over the several months in the following proportions :—

January 11 days, February 10, March 10, April 11, May 11, June 11, July 12, August 11, September 12, October 13, November 12, December 12.

This table has been calculated more particularly for the immediate neighbourhood of London; but it was believed to be very near the truth, if not absolutely so, for a very large part of the country generally. The reader will doubtless be struck by the equal distribution of the *rainy days* over each month, as shown by the above table—the greatest difference being three days only between the number for February and March, and October. Grouping these numbers according to the four seasons, we have the following result :—Spring has an average of 10·7 days in each month on which some rain falls; summer, 11·3; autumn, 12·3; and winter, 11·0. Autumn gives the greatest number of rainy days, as it also gives the largest amount of rain.

But all this seems to be rather a thing of the past. According to Mr. Symons's 'Rainfall Circular,' the number of "rainy days," or days on which rain, snow, or hail is recorded to the extent of at least one-hundredth of an inch, is very irregular over the country, and generally

[1] Williams, 'Climate of England.'

much greater than the above, at all events for last year. Still, however, it may be useful in each locality to consider the number of rainy days in a *five-year* period, and so discover if there be anything like regularity in the recurrence, and thereby know the number of rainy days that are to follow at any time or season. Of course, the average rainfall of a locality is a serious consideration in taking a farm, and the continuous record of the rain-gauge must enter into the prospect of coming harvests. The possession and *proper* use of the rain-gauge must, therefore, be considered indispensable to farmers.

The quantity of rain that falls in a given place is proportional to the elevation of the temperature; hence summer rains are more copious than winter rains, although we may have more rainy *days* in the latter.

The quantity of rain that falls in a year diminishes from the equator towards the poles. The higher the temperature the greater the difficulty of the precipitation of aqueous vapour; but, as there exists, at the same time, a much greater quantity in the atmosphere, more must be precipitated, when the circumstances that determine rain are combined. In the northern regions, where the low temperature very easily produces visible vapour, snows are very common, and the atmosphere is frequently in the condition favourable to rain.

The sum of observations made hitherto proves that the *annual* quantity of rain that falls in a country is almost always constant, so that from the greatest irregularities clear and precise inferences may be deduced. If a season has been extraordinarily wet, we may infer a drought to follow. The following table gives the rainfall of all Europe in *millimetres*, according to Becquerel. To get inches, multiply by ·004.

MEAN QUANTITY OF RAIN.

Countries.	Winter.	Spring.	Summer.	Autumn.	Year.
	Millim	Millim.	Millim.	Millim.	Millim.
England (west)	0·240	0·171	0·222	0·283	0·916
Western coasts of Europe .	0·186	0·141	0·170	0·246	0·743
England (east)	0·167	0·145	0·171	0·204	0·687
Southern France and Italy south of the Apennines .	0·195	0·194	0·133	0·292	0·814
Northern France and Germany	0·126	0·148	0·230	0·174	0·678
Italy north of Apennines .	0·139	0·139	0·276	0·354	0·908
Scandinavia	0·181	0·076	0·171	0·148	0·476
Russia	0·040	0·600	0·166	0·098	0·904

We may remark the predominance of autumn rains over summer rains in all the regions situated on the borders of the Mediterranean and the west of the Continent, up to the latitude of England. To the north and west of this band, the maximum of rains falls in summer. Thus, in the band of countries with autumn rains we find all England, the western coasts of the Continent as far as Normandy, Southern France, Italy, Greece, Asia Minor, Syria, Egypt, Barbary, Madeira. The band of summer rains comprises Northern France, Germany, the coasts of the ocean, starting from England,—the interposition of this island between the direction of the rainy winds and the Low Countries transforming them into continental countries,—in a word, all that is situated north of the central plateau of Europe, prolonged from the Alps towards the Carpathian Mountains, leaving to the south the valley of the Danube below Vienna.[1]

The preceding table shows that the annual quantity of rain which falls on the soil in Europe goes on diminishing from the south to the north, which proves the exhaustion of the vapours in proportion to the advance

[1] Gasparin, *apud* Becquerel.

H

towards the north. This effect is very striking from one end of France to the other; for, whilst the annual rainfall in the south is 0·804 m., it is only 0·678 m. in the north. For the same reason, this same quantity diminishes as we advance from the coasts into the interior of the continent. The west coasts of England receive 0·916 m., and the east coasts only receive 0·687 m. At Rochelle, the annual rainfall is 0·652 m., and at Paris it is only 0·570 m. At Nantes, the annual rainfall is 1·292 m., and Nantes is, accordingly, the most rainy place in continental Europe, although less so than Seathwaite, in England, and that to an enormous extent, its rainfall being 138·46 inches, or more than three times as much; so that Seathwaite is the wettest spot in Europe. The *driest* spots in England are Lincoln and Grantham.

The frequency of rains is an important consideration in regard to all countries. The number of rainy days in all countries is not always in relation with the quantity of water that falls, for, according to circumstances, in the same space of time the rain is more or less abundant. M. Gasparin has furnished the following table of the rainy days in all Europe, in the various seasons:—

Countries.	Winter.	Spring.	Summer.	Autumn.	Total.
England (west)	43·1	37·6	33·9	44·9	159·5
England (east)	40·0	39·5	34·4	38·8	152·7
West coasts of Europe . .	34·4	34·4	32·9	38·0	139·7
France and South Italy .	25·4	25·2	15·2	25·4	91·2
Italy north of Apennines .	25·4	27·1	25·1	26·6	104·2
Northern France and Germany.	36·1	37·0	36·8	35·0	144·9
Scandinavia	35·2	30·8	26·6	35·1	133·2
Russia	23·1	23·4	27·9	26·5	100·9

This table shows that it is in the south of France and in the south of the Apennines, and in Italy that it rains less frequently ; next comes Russia, then Italy north of the Apennines; finally, it is in England, and especially in the west of England, that it rains most frequently.

It is generally during winter, as before observed, that it rains most frequently; it rains almost as frequently in autumn as in spring; and it is in summer that it rains less, excepting in Russia.

The quantity of water that annually falls, both in rain and snow, upon the surface of the earth, is much less considerable than we are apt to imagine from the number of rainy days, and at the sight of those downpours which sometimes last for entire days. The diluvian rains of the tropics do not give more than 0·2 m., or less than 8 inches in twelve hours. In our latitudes, when the quantity of water falling in a single day exceeds 0·03 m., or 1·17 inch, the plains are soon inundated.

In general, the quantity of rain that falls annually diminishes as we recede from the sea. On the west coast of England, the rainfall is 0·92 m., whilst it is only 0·65 m. in the interior of the country. On the coasts of France and Holland, the rainfall is 0·68 m. and in the interior 0·65 m. In the plains of Germany, it is only 0·54 m., and at Baden, from 0·43 m. to 0·46 m.

Supposing the atmosphere divided into slices of 860 *metres* (say, *yards*), whose temperature should go on decreasing at the rate of 5° from 86° to 14° Fahr., M. Becquerel has calculated that the total quantity of water in vapour in the atmosphere at any instant, does not exceed the weight of a layer of water 0·1 m. (or less than 4 inches) in thickness, covering the entire globe! This explains the small depth of the layer of water which falls, even in the most tremendous downpours.

Hence, the *miraculous* nature of the rain which caused the Deluge described in the Bible. It would be a tremendous shower which should fall at the rate of two-hundredths of an inch per minute (1·20 inch per hour); but the rain of the Deluge must have fallen at the rate of more than five inches per minute, nearly 318 inches per hour, 7626¾ inches per diem, and the total rainfall of the 40 days and 40 nights must have been 305,070 inches, or 25,422½ feet, over 4¾ miles.[1]

There is another point of some importance, namely, the quantity of rain as influenced by the relative height of localities.

Although it rains much oftener on mountains than on the plains at their feet, the quantity of water that falls each year is notably more considerable on the plains than on the mountains. This fact is perfectly established, and we can explain it by another, namely, the precipitation of the vapour of the lower layers by the cold water coming from above. But the difference is so great that we must necessarily admit the presence of local influences, generally unknown. At the Observatory of Paris there are two rain-gauges, one on the ground in the courtyard, and the other on the terrace, twenty-seven metres higher. The first receives, in the

---

[1] The calculation is as follows :—(1) It rained 40 days and 40 nights. (2) The water stood 15 cubits above the mountains : 15 cubits = 270 inches (15 × 18). (3) The highest peak of the Himalayas, 25,400 feet = 304,800 inches,—to which add (2) 270 inches, making 305,070 inches of rainfall,—which, being divided by 40, gives 7626·75 inches per diem ; 317·82 inches per hour ; 5·29 inches per minute.

annual mean, 0·57 m., whilst the second only receives
0·50 m.; the difference, 0·07 m., cannot be entirely attributed to the small difference in the level of the two
rain-gauges.   However, the subject is important in
considering the rainfall of localities, in which the relative
elevation should always be taken into account.

In the present method of ascertaining rainfall it appears that the estimates must be considered mere approximations, although sufficient for practical inferences. Rain-
gauges are made of every diameter, and placed in every
variety of position, and at various elevations.   All these
circumstances more or less influence the results obtained.
But we apprehend that the great objection to all rain-
gauges is their fixed vertical or upright position (when
projecting above the soil), the consequence of which is
that they can only receive vertical rain completely;
whereas by far the great majority of rains come in a
slanting direction, indeed, the direction of the fall may
be often observed to vary incessantly, even during settled rains.   Such being the case, it is evident that only
isolated rain-gauges (that is, fixed in a trench with their
mouths on a level with the soil) or *tilted-funnelled*
gauges should be used. We feel sure that no other kind
can be accurate.   Of the latter kind, a gauge was introduced by Professor Phillips having one horizontal and
four vertical funnels, facing E., W., N., S.   Absolutely
vertical rain enters the horizontal funnel; absolutely
horizontal—say from due east—enters the funnel facing
east; if at any intermediate angle, it will partly fall into
two or three funnels, each being provided with separate
pipes and taps; the quantity caught by each is known,
and the angle and point whence the rain fell can be
easily calculated.

Mr. Symons has improved this principle in a gauge the tilt of which is not fixed, but varies with the pressure of the wind. In a dead calm, the funnel is horizontal, and in a gale it will be tilted to an angle of 70° or 80°, thus always keeping at right angles to the wind, and catching more rain than any other in windy weather. We do not hesitate to say that if all the gauges in the country were of this description or isolated, a very considerable difference in the rainfall estimates would be the result.

With respect to the diameter of the rain-gauge, it seems that it should not be *less* than five inches; and the edges should be sharp, and not bevelled off, as in some cases to an inch or two, against which the inclined rain-drops strike, are reflected in the usual way, and fall into the gauge, without having any right to be there in the estimate.

The following table shows the different results from rain-gauges according to their elevation above the soil, during nine months.

	Total inches.	Excess above fall at one foot.
Gauge at 1 foot above ground	23·08	0·00
Gauge at 6 inches above ground	23·24	+0·16
Gauge at 2 inches above ground	23·61	+0·53
Gauge level with the ground	23·91	+0·83
Gauge *isolated* (in a pit)	23·60	+0·52 [1]

Doubtless, the isolated gauge gave the only correct result. The one on a level with the ground probably included *splashings* as well as rain-drops. According to Mr. Symons, there appears to be a considerable increase in the amount collected by gauges whose funnels are of glass and pot, that is, with unoxidizable surfaces, the excess being 6 per cent.; and gauges thoroughly pro-

[1] Symons, 'British Rainfall,' 1865.

tected from evaporation may exceed unprotected ones by about the same amount. Therefore, a thoroughly protected gauge, with a permanently smooth surface, would apparently exceed an ordinary one by nearly 12 per cent. —an enormous difference, and one which needs close scrutiny, to which Mr. Symons has now submitted them.*

As to locality, the rain-gauge should be as exposed as possible to all the points of the compass,—at any rate in a south aspect, and never in a north aspect, where its record will infallibly be too low, the majority of our rains coming from the southward and westward. The instrument should be as close as possible to the surface without running the risk of receiving splashes from the ground; probably six inches will suffice. The higher the gauge is fixed, excepting in extraordinary circumstances (of which more in the sequel, Chapter XXVI.), the less the quantity of water it receives.

The results of snow must, of course, be added to the year's rainfall. Herein, obviously, the rain-gauge is almost useless. Hitherto the method employed has been to measure a portion of snow in the funnel of the rain-gauge, to melt it down at a "moderate temperature," and then to measure the water as usual. But melting snow evaporates; we cannot always get the snow at a uniform density by packing it in a funnel; and, besides the evident uncertainty, the process is decidedly cumbersome. A snow-gauge is wanted, and we have endeavoured to contrive one, the principle of which is based on existing data as to the relative quantity of water corresponding to the various depths of snow.

* The result of this scrutiny will be shortly known, and must be of great interest to observers.

As the ratio depends entirely on the density of the snow measured and melted, it is not surprising that it differs exceedingly in the results of manipulators,— indeed, as Mr. Symons states, from one-thirty-fifth to one-fifth, through almost every intermediate figure. We cannot help thinking that this great discrepancy results mainly from the different modes of effecting the same purpose, differences in the density of the snow by packing (as we have seen the operation described), differences of temperature in melting, and consequent loss by evaporation. However, of twenty-six measurements by various manipulators, given by Mr. Symons, the mean is 14 inches of snow, equal to 1 inch of rain.

SNOW MEASUREMENT.

Depth of Snow.	Yield in Water.	Depth of Snow to yield 1 in. of Water.	Observer.	Depth of Snow.	Yield in Water.	Depth of Snow to yield 1 in. of Water.	Observer.
in.	in.	in.		in.	in.	in.	
1·70	·048	35	G. J. S.	4·50	·366	12	E. J. L.
·75	·0. 0	25	M. F. W.	·50	·042	12	G. J. S.
3·00	·140	21	E. J. L.	6·50	·527	12	E. J. L.
1·00	·048	21	G. J. S.	·10	·009	11	G. J. S.
1·00	·049	20	G. J. S.	2·12	·194	11	G. J. S.
1·00	·052	19	M. F. W.	3·00	·300	10	M. F. W.
18·00	·950	19	H. S. E.	6·00	·587	10	M. F. W.
2·00	·112	18	G. J. S.	12·00	1·270	9	H. S. E.
1·50	·088	17	M. F. W.	3·50	·391	9	E. J. L.
1·75	·120	15	M. F. W.	·50	·090	6	G. J. S.
1·00	·068	15	M. F. W.	2·00	·370	5	W. F. H.
·25	·050	15	G. J. S.	·50	·100	5	G. J. S.
·15	·012	13	G. J. S.	5·50	1·215	5	M. F. W.

Mean of the above 26 measurements :—14 inches of snow = 1 inch of rain.

Observers.—H. S. Eaton, Little Bridy; W. F. Harrison, Weybridge Heath; E. J. Lowe, Nottingham; G. J. Symons, Camden Town; Col. M. F. Ward, Calne.

Mr. Symons is, as everybody knows, a careful and reliable authority, and, notwithstanding the above result and the experiment of Mr. Harrison, of Bartropps, Weybridge Heath, giving the ratio as 5·4 of snow to 1 inch, he persists in fixing the ratio as *one-twelfth,* or 12 inches of snow to 1 inch of rain.[1] Colonel Sir H. James gives the same from results in Canada.

Taking, then, 12 in. snow as equal to ·96 in. water, we have contrived the following snow gauge—

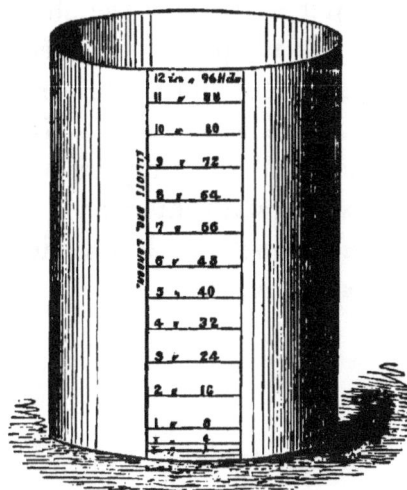

STEINMETZ'S REDUCTION SNOW GAUGE.[2]

The only peculiarity in the snow-gauge is the graduated glass showing the quantity of rainfall corresponding to the existing depth of snowfall. As the thing is experimental, the graduated glass is not a fixture, but merely slides down its supports, so that should observers fix upon any other ratio, a glass slide with the requisite graduation can be furnished by the makers. On the left is the depth of snow, on the right, the corresponding rainfall to be added to the register as usual, down to the hundredth. Should the snow be inclined to one side of

[1] 'British Rainfall,' 1865.
[2] Made by Messrs. Elliott, Brothers, 449, Strand.

the gauge more than the other, it may be gently brushed into the horizontal before making the observation.

Obviously, the same rule should apply to the position of the snow-gauge as to that of the rain-gauge; for elevation will make a difference in the quantity of snow received, and an isolated snow-gauge, level with the ground, will be required to show the absolute quantity fallen at any time. If not perfectly exposed to all points of the compass, preference must be given to a northern and eastern aspect for a snow-gauge. It should be remembered that snow undergoes rapid changes on alighting, among the rest, crystallization and evaporation, of which more in the sequel, pp. 203–4; it should therefore be measured speedily.

CHAPTER VIII.

WET AND DRY YEARS.

If the problem of particular wet and dry years be still unsolved, still a few facts have been established respecting the predominance of wet or dry ones during the decades of the last fifty years, and it seems possible to reason out a sort of guidance for our future expectations from the considerations in question.

It is a curious coincidence that the summer of 1766 was similar to the last; the Earl of Chesterfield, in one of his letters, says, "there has been no summer as wet as this within the memory of man; since March we have not had one day without rain, until August 1st, 1766." The year was one of almost incessant rain, so much so, indeed, that "on the 2nd of June the hay-makers assembled at the Royal Exchange, to the number of 440, when a collection was made for them on account of the heavy rains, which prevented them from working."[1]

[1] Among the records of the City relative to the assize of bread, published in 1766, under the authority of the Lord Mayor, the following entry appears :—" October 7, 1766. By the assize of bread

The proper exposition of the inquiry seems to require
the preliminary consideration of the sources of our rain
and their causes.

Drought and sunshine in one part of Europe may be
as necessary to the production of a wet season in another,
as they are found to be on the great scale of the continents of Africa and South America, where the plains,
during one half of the year, are burnt up to feed the
springs of the mountains, which, in their turn, contribute to inundate the fertile valleys, and prepare them
for a luxuriant vegetation.   During the wet summer of
1816 in England, the middle of Europe was subjected to
excessive rains, at the same time that the north, or the
parts east of the Baltic, about Dantzic and Riga, were
suffering from drought, and in all probability furnishing
the water in the shape of vapour.

In the spring and summer, as Luke Howard observed,
both the direction of the winds and the relative state of
temperature seem to forbid our receiving much rain
from the Atlantic.   It is the south-east wind which is
so intimately connected with electrical manifestations,
with hail, rain, and thunder.   Now, vapour brought to
us by such a wind must have been generated in countries lying to the south and east of our island.   It is
therefore probably in the extensive valleys watered by
the Meuse, the Moselle, and the Rhine, if not from the
more distant Elbe, the Oder, and the Weser, that the
water rises, in the midst of sunshine, which is soon
afterwards to form *our* clouds, and pour down in *our*
thunder-showers.   England   probably   does   the   same

this day, the peck loaf to weigh 17 lb. 6 oz.; wheaten bread 2*s.* 8*d.*,
household 2*s.*; average price per quartern loaf 8*d.*"   On the 29th of
September an embargo was laid on the exportation of corn.

office for Ireland,—nay, the eastern for the western counties of South Britain.

It is otherwise in winter. During winter, when the surface of the ocean is giving out heat to the air, it may be supposed also to give out vapour in greater quantities than the temperature of the air is prepared to retain. Hence, the Atlantic, during the winter months, or rather the interval between the autumnal equinox and the winter solstice, is probably the great source of our rains. The impetuous gales which, at this season, move over its surface and impinge upon our western shores, may possibly bring us much vapour from the superior atmosphere of the tropics in which they originate.

Such seems to be the theory of our rains : the vapour generated in our own latitudes is condensed into rain by the south-west winds, which are *cooler* from blowing over the ocean in summer, and during the winter the same winds come to us charged with vapour from the warmer ocean, to be condensed by a lower temperature ; the entire theory of rain being the mixture or brushing of different currents of air saturated with moisture of different temperatures.

Now, let us examine the general results of this state of things with respect to the wet years during the last fifty years, namely, from 1815 to 1864, by means of the Table drawn up by Mr. Symons.

Anything like absolute periodicity it is impossible to perceive; but, in the absence of this, we find something like a substitute for it in the progression of wet years throughout the entire period, from what may be taken as a maximum in the first decade, down to a minimum in the last.

In the first ten years (1815–1824) the rainfall of

seven years was above the average; in the next ten
(1825–1834) six; in the next (1835–1844) five; in the
next (1845–1854) four; in the last (1855–1864) three.

This regular progression is remarkable, and has all
the appearance of a law; and as three wet years must
be considered the smallest number possible or probable
in a decade, possibly we may conclude that the last decade concluded a period of minima—characterized by
excessive drought—and that we are now commencing a
decade which will ultimately exhibit a maximum, or, as
it may be, an increasing progression in the number of
wet years towards the maximum of the first decade of
the previous period.

Another fact is apparent from this record : out of the
first twenty-five years of the fifty years' period, sixteen
were above the average, and in the next twenty-five
years, sixteen were below it.

There is also an ominous ten-year period, or something very like it, corresponding with full, heavy, or
excessive rainfall in Great Britain. Thus, in 1825, the
rainfall was 26·57 inches; in 1836, 33·49 inches; in
1846, 29·57 inches. In the year 1856, it was a full average, after two years below the average; but now, in 1866,
it is a maximum which is likely to equal the highest.

We need scarcely state that all years of excessive rainfall have been followed by bad harvests. Thus, in 1836
(rainfall 33·49) there was a bad harvest followed by three
bad harvests in succession; in 1846 (when the Corn
Laws were repealed) the rainfall was 29·57, and there
was a bad harvest, followed by another very bad harvest.

Fortunately, however, the excessive drainage and
clearance of the country will, in future, tend to prevent
the natural consequences of excessive rainfall, and a wet

year may now actually *produce* a good subsequent harvest, by supplying the requisite moisture to the soil.
But, it must be remembered that excessive drainage
renders us liable to frequent droughts, necessarily followed by excessive downpours by way of compensation.

We now come to the important part of the inquiry
relating to the distribution of rainfall over the country—
the regions of excess or deficiency.

During the last period of ten years (1855–1864) there
was a decrease of rainfall averaging four per cent. over
the entire British Isles, but unequally distributed, the decrease being exchanged for an increase in parts of Ireland
and the south of Scotland.   In England the amount of
decrease varied up to eighteen per cent., and was never
below two per cent.

Although at first the figures seemed very discordant,
yet on drawing lines on a map of the country, some
order seems to become evident, namely, that the maximum deficiency existed along a line running nearly
south-west towards north-east—from Cornwall to the
Wash—along which line the present year's rainfall is
already greatly in excess.   This is, therefore, a compensation which may always, accordingly, be sooner or later
expected in districts visited by continuous drought or
deficiency of rainfall.

Proceeding north-westward the deficiency became less,
until the parallel line running through the centre of
Ireland and passing into the North Sea at Edinburgh,
marked a district in which no deficiency existed, but, on
the contrary, an excess of nearly ten per cent.   This is
the main line of the country's rainfall, one part of it,
Seathwaite, as before stated, being the wettest spot in all
Europe, having an annual rainfall of 138·46 inches, and

having received during the last September an excess of 8·14 inches above its enormous average.

The next districts follow nearer to each other, and seem to involve the eventual adoption of west to east, instead of south-west to north-east; possibly this is not really the case, as we suspect, but due to errors of observation at the lighthouses, whence most of the values assigned for Scotland are derived, or it may arise from the modifying influence of Ireland not being felt in those higher latitudes.

As a general proposition the full, average, or excessive rainfall seems to be intimately connected with the nature of the country on the line, as to tall vegetation, trees, and elevated points,—in a word, a region where man has not as yet shown himself as the uncompromising leveller in creation. On the other hand, the deficiency seems in some degree connected with the large drainage operations in the midland and eastern counties of England.

Drainage influences rainfall in two ways. It rapidly carries off the water of rains and thus diminishes the supply for evaporation, which is the supply of future rain. In the next place, by carrying off the moisture, it raises the *temperature* of a country or district, and thereby interferes with the condensation of vapour resulting from a lowering of the temperature.

These facts should be turned to account by all who are engaged in agricultural pursuits. It is evident that they may guide us in the cultivation of the soil, in the selection of localities for farming, in the choice of crops,—and generally in adopting localities with reference to their adaptation to the requirements of health or the human constitution.

Finally, if, as it appears, a progression of wet years characterizes successive decades, it seems that we should turn our attention to the *green crops* by way of compensation in such decades as that in which we appear to be, rather than trust to the chance of a bad harvest,—especially with the fact that bread will never again be dear, and meat will never be cheap.

There was a time when England was a country of forests,—when, like America and continental Europe of the present day, the old woods occupied large tracts in every part, as the domiciles of wolves, wild boars, and other beasts of prey, which roamed the gloomy solitudes, occasionally emerging to the open glades to pick up a juvenile cottager, and carry him off to feed its young cubs. These forests were the parents of marshes and bogs, which constituted the never-failing feeders of the numerous rivers and streams that intersected the land in every direction. In those early times, the alternating seasons were more marked and regular than since cultivation has covered the land with its vegetable productions. The summers were more intensely hot; the winters more piercingly cold; the spring more regular in its appearance, and more to be depended upon for its genial and renovating influence; and the autumn more settled in its weather, for the work of gathering in the fruits of the fields. Such is the account given by chroniclers; and we will at once give them credit for correctness and veracity, though perhaps not to the full extent of their representations. Certain it is, that the clearance of the forests, and the consequent drying up or draining of the marshes and bogs, have caused a material alteration not only in the entire face of the country, but in the supply of water—the rivers formerly de-

rived from those reservoirs, and also in the periodical
amount of rainfall and the regularity of its distribution.
Many streams throughout the country, which formerly
supplied large mills with unfailing water-power, except
in the very driest of seasons, are now, with vastly improved machinery requiring less power, frequently unable to work; and almost all are compelled to be sup-
plemented by steam-power to make good the deficiency
in the rainfall. In addition to this cause of the declining water-power of the country, we may instance also
the rapidity with which the water that falls is hurried
to the ocean,—by the system of drainage throughout
the country, by which the land is cleared of rain almost
as soon as it falls, having now none of the ancient reservoirs, in the shape of bogs and marshes, to receive
and retain it for future use. All the old water-millers
complain that the rainfall, however heavy, does them no
permanent good. It just occasions a momentary "flush,"
which is rather injurious than otherwise, being in excess
of the requisite power, instead of being held in reserve
by the marshes, and, above all, by the *subsoil* of the
adjacent lands. This was formerly, in the undrained
state of the country, perhaps the largest source of supply,
because it extended over the whole area, and yielded its
reserve deliberately and in driblets; whereas now, the
shower, as soon as it falls, finds its way into the drains,
and from thence to the river, in a few hours; causing,
as we have stated, a temporary flush of water, but of no
permanent benefit to the miller, who, in fact, is compelled to stop his mill for the time, and let the precious
fluid run past by the back-water course.

The proportion of forest or woodland required for an
agricultural country, in order to ensure a regular and

sufficient rainfall without violent storms, is estimated by
Rentzsch at 23 per cent. for the interior, and 20 per
cent. near the coast. This estimate relates to Germany,
but in England, the proportion, according to the same
authority, is only 5 per cent., and even this is reduced
by Sir Henry James, the head of the Ordnance Survey
Department, and who ought to know, to 2·5 per cent.
This is certainly a very small proportion, and below that
of every other country, the next lowest being Portugal,
which has very little woodland. For fuel, England at
present requires no wood, whatever she may do a hundred or a thousand years hence ; but for the purpose of
ensuring a regular supply of rainfall, and reducing the
force and number of violent storms, the planting of the
waste and poor lands would undoubtedly be beneficial,
although, from the large proportion of seacoast, a less extent of forest is required than on the Continent. An instance from Boussingault is quoted by Professor Murray.
"The Wolf-spring, in the commune of Soubey, furnishes
a remarkable instance of the influence of woods upon
fountains. A few years ago, this spring did not exist.
At the place where it now rises, a small thread of water
was observed, after very long rains, but the stream disappeared with the rain. The spot is in the middle of a
very steep pasture, inclining to the south. Eighty years
ago, the owner of the land perceiving that some firs were
shooting up in the upper part of it, determined to let
them grow, and they soon formed a flourishing grove.
As soon as they were well grown, a fine spring appeared
in place of the occasional rill, and furnished abundant
water in the longest drought. For forty or fifty years,
this spring was considered the best in Clos-du-Doubs.
A few years since, the grove was felled, and the ground

turned again to a pasture. The spring disappeared with
the wood, and is now as dry as it was ninety years
ago."[1]

In the previous chapter we pointed out the evil consequences to the climate, as far as the fertility of the soil
is concerned, resulting from the clearance of forests,
which are one of the means provided by Nature for
securing adequate rainfall. They are not the only cause
of rain, as must be obvious, for rain mainly depends
upon atmospheric conditions. But whilst forests tend
to secure an adequate rainfall, they also constitute a
natural security against inundations from excessive rainfall, by preventing the crumbling down of the mountain-
sides in landslips and ravines, the ravaging of the plains
by floods carrying down vast masses of stone and gravel,
all which results from the binding roots of forest trees.
Trees also prevent the freshets, almost invariably attending of late every spring thaw and every autumn rain.
On the Continent, the suddenness and rapidity of the
flow of waters have been determined by baring the hills,
by widening the valleys, and smoothing down the
grooves through which paths were opened for the slide
of the felled timber. As a writer in the 'Times' puts
it, "Something like a universal loosening of the cata-
racts of the mountains has been effected; and the fall of
the streams, from their sources to their mouths, has
been increased in speed, in proportion as the filling up
of their channels, by the accumulation of their deposits,
has clogged and slackened their course in the plain.
The almost total disappearance of green vegetation in
the higher mountains, and the overthrow of the barrier
that dense and lofty timber raised against the violence

[1] The 'Mark Lane Express.'

of the winds, have both affected the degree of moisture and the heat in the atmosphere, and substituted an alternative of long droughts and sudden storms, instead of the mild and gradual transitions by which season followed season in those genial climates. Atmospheric influences are certainly not amenable to calculation by human rule and compass. The science which reasonably aims at determining their phenomena is still in its infancy; but there is little doubt that much of the suffering inflicted on mankind by what was attributed to the disorder of the seasons, must be accounted for by their own recklessness and improvidence; and that, in many cases, as in these matters of inundations, what was in nature only a casual and exceptional evil, has become permanent, and been greatly aggravated by circumstances under man's free control."

There is still a third hint that we may take, if we like, from these deluging rains. Why do we not take means to preserve the flood-waters? Why has not every district and town its well-made, well-kept reservoir, always ready to prevent the ravages of inundations, and to afford a cheap and plentiful supply of the fluid which becomes so precious in the time of drought? The present system of water companies is to intercept the regular flow of streams, and reject the flood-waters. Nothing can be more wasteful. Again, some of the companies draw upon the natural lakes, which must eventually be exhausted, unless " heads" are raised at their outlet, to retain a quantity from the autumn and winter rains equivalent to a year's demand.

All our companies should be compelled by Parliament to draw their supplies from artificial lakes. All our manufacturing towns are situated within a moderate distance from mountain ranges, where lakes, ranging from six to eight miles long, could be formed by a "head" of from 100 to 150 feet high. The cost of these heads, according to a 'Times' correspondent, would not probably exceed £1,000,000 or £1,500,000, and the aqueduct not more than £20,000 per mile,—small sums when the importance of the subject is considered. The Metropolis is large enough to afford a larger cost of conveyance. The gigantic tanks artificially formed in Ceylon would give our hydraulic engineers some idea of what will shortly be required for our increasing population. But a proper method should be adopted in constructing dam-heads. The bursting of the Holmfirth and Sheffield reservoirs was really owing to want of proper consolidation of the head. The modern way of forming dam-heads is by tipping waggon-loads of earth weighing not less than half a ton, generally a ton. Of course, these large masses of earth take years to consolidate. The earth should be carried in baskets upon the head, the gangs of carriers to be so arranged that each quantity, after it is thrown down, should be trodden upon by the whole gang. This may appear, at first sight, a very expensive process; but it has been found the cheapest and most expeditious way of coaling the mail-packets in the West Indies, where from fifty to one hundred tons per hour are put on board the mail boats, carried on the heads of the "coloured ladies," who often continue the work through the whole night. At any rate, means must be taken to consolidate the mass more effectually than at present.

Such are the hints suggested to us by the rainfall and floods of September, 1866, and we apprehend that their importance will recommend them to all concerned, if not to the nation at large, which is certainly very deeply interested in the matter in all its bearings.[1]

[1] " If the farmer would be content to deal with his grain as he does with his other fruit, to take the trouble of a double ingathering, to crop ears of corn as he would basket cherries or bin hops, and then to mow his straw as he mows his hay, he would incur a first expense perhaps double of his present. If, when the corn arrived at a proper state of maturity, irrespective of barometer or weathercock, he were to turn all available hands, women and children more profitably than men, on his well-eared fields, each provided with a basket such as that in which seed-corn is carried, slung round the neck, and a small knife, or a triangular piece of iron, worn like a tailor's thimble on the forefinger, the whole of the corn would be rapidly secured, being plucked or rather bent over the knife immediately beneath the ear, dropped at once into the basket, and never suffered to touch the ground, nor knocked about to the loss of a single grain, before it was safely deposited in the granary ; then, if wet, it could be readily and inexpensively subjected to a simple process of drying—a process which, if once adopted, would be likely to become universal, whether the corn were ingathered wet or dry, as it would entirely obviate the painful and clumsy process of thrashing. If the ears of corn were placed in baskets or wire trays, tier above tier, so that a strong and sustained blast of air were driven through them, whether set in motion by a fan or by the draught of a lofty chimney,— the great power of which is well known to those who have had to deal with really high stacks, —not only would all superfluous moisture be rapidly absorbed, but the ear would become so dry as to cease to retain the grain, which would fall from the basket, as from a sieve, on the slightest motion. The precious part of the crop thus secured, the farmer would take his own time for mowing the straw, or cutting it by machinery ; at least he would so do until practice in drying the wheat itself should teach him that it was a false economy to lose either straw or hay in unfavourable weather for want of proper arrangements for drying these valuable crops."—*Builder.*

CHAPTER IX.

OF TEMPERATURE AND MOISTURE IN THE MAT-
TER OF HEALTH IN PLANTS AND ANIMALS.

IN relation to atmospheric temperature as shown by the
thermometer, and atmospheric moisture as indicated by
the hygrometer, meteorology promises to throw light on
the animal functions under different conditions in health
and disease.

Already are physicians beginning to have more faith
in the thermometer than the feeling of the pulse, which
was long ago declared to be a most fallacious thing—*res
fallacissima, pulsus.* By means of a thermometer,
whose bulb is placed in the armpit, we may now detect
the march of disease, it having been established that
every disease is characterized in its various stages by a
particular or definite temperature.

In the weekly records of the Registrar-General, the
existing temperature and the daily range of the thermometer are intimately connected with the Bill of Mortality. The meteorological discovery of there being a
difference of a few degrees in the mean temperature
of the week, explains the otherwise mysterious work of
death in the sick-chamber. Man can bear almost any

amount of heat, but the slightest lowering of temperature tells more or less on the human machine so wonderfully made. The dry east wind kills off its hundreds,
and the moist south-western agitates the asthmatic and
those whose "troublesome cough" makes them dread
the autumn. With all these vital matters meteorology
is specially concerned, and its development is destined
to prove of incalculable benefit to mankind, not only in
its relation with the coming storm, which it will be able
infallibly to predict, the expectancies of the seasons, and
the changes of the weather from day to day, but also in
its relation to the more hidden causes and influences
with which the health of nations is concerned.

The cause of summer diarrhœa, or *cholerine*, for example, used to be attributed to irregularities of diet
during its prevalence; but the predominance of this
complaint among the infantile population seemed to
throw doubts on this conclusion, and, indeed, to render
that cause totally inadmissible. Now, however, meteorology suggests what seems to be the true cause of this
ever-recurring visitation. The increase in the number
of the subjects of this disease as the summer temperature
rises, and the decline with the colder season of the year,
seem to prove that *temperature* is the exciting cause of
this abnormal condition of the system. From a table
drawn up by Dr. Dundas Thomson, the President of the
Meteorological Society, it appears that there is a gradual increase of diarrhœa from the early months of the
year up to June, at which period the number of cases
may be about 500, when it advances in intensity, so
that in July, the number of cases is 2200. That is its
maximum, whence it descends in August to 1900, in
September to 700, and in the remaining months settles

down to about the same minimum as in the early months of the year; and such is the march of the *temperature* throughout the months of the year. The excess of accumulated heat in June is double that in May, and the cases rise from 160 to 544. Again, in July, the excess of accumulated heat is nearly four days above that of June, and the cases increase to 2293, while in August they fall to 1934, the excess of accumulated heat having diminished by nearly two days. We cannot expect an exact numerical relation between the heat and the disease, because the cases are limited, and the influence of the heat is cumulative from one month to another, which, of course, cannot be precisely estimated when acting on vital beings endowed with a nervous resisting-power varying in each individual.

One of the most interesting facts in confirmation of this view of atmospheric temperature upon the human system, and of the disturbance of the normal diffusion of the animal fluids, is the established conclusion which has been arrived at in Calcutta,—that the hot season is the least favourable to vaccination. In 10,102 cases vaccinated in the cold season (mean temperature $= 75\cdot6^\circ$) in that city, from the 1st of November, 1853, to the 31st of March, 1854, 96·07 per cent. were successful, 1·67 were partly successful, and 2·25 per cent. failed. On the other hand, in the hot season, with a mean temperature of 86·5°, between the 1st of April and the 30th of September, 1854, 2100 were vaccinated; of these only 73·76 per cent. were successful, 4·90 per cent. partially, and 21·33 per cent. were failures.

These considerations, therefore, show the important bearing of meteorology on health, and how much of the animal nature is dependent on purely physical conditions.

In this country, during hot weather, the direction of the fluids of the body is disturbed in those affected with diarrhœa. The fluids which, in the healthy state, pass from the intestinal canal into the blood, have, in the abnormal state produced by heat, their action reversed, and then they pass from the blood into the intestinal canal.[1] With regard to extraordinary epidemics, it seems that every country has a standard climate, depending on latitude and geographical position, and any marked deviation from that standard is almost certainly followed by a corresponding increase of sickness. Hence, the great account to which meteorological observations may be turned by the community by being forewarned of the impending outbreak.

It appears that wet seasons are by no means the unhealthiest. The highest death-rate of twelve years, 23·9, occurred with the smallest rainfall, 16·7 in., in 1864; and the lowest rate, 21·2, in 1860, with the heaviest rainfall, 32 in., in 1860. This may doubtless be accounted for in many ways, and principally by the cleansing influence of the rain during the summer upon the impurities of towns, which in dry weather proves so noxious in crowded populations; but it is also very possible that the greater humidity of the air produced by the rain, may be beneficial to all persons suffering from affections of the lungs. A heavy fall of rain has frequently stopped, for a time, the development of Yellow Fever in Jamaica, probably by mechanically preventing emanations from the earth. We may be over-draining and over-drying our land and our air, to the detriment of both animal and vegetable life; we may have too deep-

---

[1] 'Proceedings of the British Meteorological Society,' vol. ii., President's Address.

rooted a dread of a pond near the house, if it be kept
clean, and also of trees, which may serve the double
purpose of shelter, and of preventing too complete an
evaporation from the ground. It also appears obvious
that the advantage of shelter for a house, especially from
the east, is now often too much overlooked in the desire
to build on high and dry situations. We must not forget the fact that a candle in a draught will always burn
wastefully, and that our bodies, after all, are but candles,
which burn economically according to their situation.

And now, as to the hygrometer, and its contribution
to the health of the community. At our watering-
places and invalid-stations the degree of dryness in the
air is an important element in the sanitary value of the
locality, and it is only by the hygrometer that the fact
can be accurately ascertained. The hygrometrical record of any such place would, if satisfactory, be its best
recommendation to the invalid ; and all who seek health
in a change of place, would do well to ascertain the condition of the locality as to humidity in the air, if that
be essentially objectionable.

In the sick-chamber the comfort of the patient greatly
depends upon the condition of the air as to moisture.
The cause of the uneasy sensations will be at once re-
vealed by the hygrometer, as, for instance, during cold
frosty weather, when the air of our rooms is often too
dry,—which will be shown by the great difference between the readings of the two bulbs. This state of
things may be easily remedied by exposing water in
shallow vessels, such as a meat dish, in order to evaporate and mix its vapour with the air in the room, thus
imparting to it the requisite amount of moisture. Of
course if the water be heated the process will be quicker ;

or a tube might be fixed to the spout of a kettle on the fire, kept boiling, which will soon supply the desirable amount of vapour.

On the other hand, if the readings of the two thermometers be alike, or very nearly so, it will be evident that the air of the room is saturated with moisture, or very nearly so. In this case the room must be made warmer, all water must be either removed or carefully covered over, and if greater dryness be still required we may place in the room such substances as have the property of rapidly absorbing water, such as lime or sulphuric acid, the latter in a strong earthen dish. In short, by such simple means, it is evident that we may produce any sort of climate that may be suitable to our invalids according to their ailments or the state of their organs and constitution, without sending them abroad or to a distance from their homes. Nay more, by such artificial means, which we can always regulate with the greatest nicety, we might produce a better climate than could be found anywhere, for in the very best of localities sudden changes are apt to spoil the effect of the best conditions of health. Dr. Mason found at Madeira, where he resided for health, that during the *leste,* or dry wind of the island, his hygrometer often indicated 24 degrees of dryness, while two hours before the setting in of the wind the dryness was only 20 degrees.

The use of the hygrometer will show why stoves, now so generally in use, are injurious to health, from the excessive dryness of the air which they produce in a room, causing moisture to evaporate too freely from the skin, with all the painful consequences to the general health of those who come under their influence. Evaporating dishes should be set up, to neutralize their effect.

A difference of from six to eight degrees between the reading of the two thermometers will be generally found to produce or to accompany a pleasant degree of humidity in a room.

In a room, the hygrometer should be placed at some distance from the fire, and not exposed to open doors or currents of air; the best place is a recess on the same side of the room as the fireplace.[1]

Of less, but still of very great importance is the use of the hygrometer in our hot-houses, greenhouses, conservatories, etc., as well as in our gardens, where our hardier vegetable pets are sometimes endangered by changes in the hygrometric condition of the air. In the latter, abundant dews are often connected with white frost, which may prove very disastrous to the plants. Now if we consult the hygrometer in the evening, and find the two bulbs pretty much alike in their reading, whilst the sky is clear, even should there be no rain, there will certainly be a copious deposit of dew. After sunset, all bodies on the surface of the earth, after having been heated by the sun during the day, radiate or give off their heat into space, get cold and soon reach the dew-point, when they become covered with moisture, provided there be no clouds or other objects to check the dispersion of heat or reflect it back to the earth. If the temperature be low, the dew becomes hoar-frost. All gardeners know that it may be produced without the thermometer falling to the freezing-point. The reason is that the dew evaporates so rapidly that it deprives itself of a considerable amount of heat, so as to congeal into minute needles the water remaining on the plants and other bodies.

[1] Glaisher, 'Hygrometrical Tables.'

But the frosts of spring and autumn, which are so frequently injurious to the crops of the farmer and gardener, proceed generally not from the congelation of moisture deposited from the atmosphere, but from the congelation of their own proper moisture by the radiation of their temperature, caused by the nocturnal radiation, which in other cases produces dew or hoar-frost. The young buds of leaves and flowers in spring, and the grain and fruit in autumn, being reduced by radiation below 32°, while the atmosphere is many degrees above that temperature, the water which forms part of their composition is frozen, and *blight* ensues. This explains the cause of the evil, and suggests the remedy. It is only necessary to shelter the object from exposure to the unclouded sky by matting, gauze, or other covering.

But this is not the only consequence of radiation. The sudden evaporation carries into the atmosphere a large quantity of aqueous vapour, and so, after a night or two of hoar-frost, we often get rain, in accordance with the common proverb, "After three nights' hoar-frost comes rain."

Now, all this may be known beforehand by carefully noting the hygrometer, and then we can prevent or diminish the deposit of dew by covering our plants with a screen, and the quantity may be lessened by placing beside them an upright board, to prevent a portion of the radiation into space.

Certain bodies get more dew than others, and the quantity received differs in all. This results from their power of radiation. The more bodies radiate heat, the more dew they make. Plants get more dew than the soil; the various kinds of soil more than the metals; sand more than hard ground; chips more than the pieces of the wood from which they are chopped.

But it is in conservatories that the hygrometer is indispensable. How often do our greenhouse plants become shrivelled or weak before we have the least sus-
picion that there is any alteration in the humidity of
the air! Then, as soon as we become alive to the fact,
we drench them with water, without taking their actual
requirements into consideration. On the other hand, if
we fancy, from our own sensations, that the air of the
conservatory is dry, we sprinkle water about, without
measure. But, as Mr. Glaisher observes, our sensations, with regard to heat and humidity, are very fallacious guides. Every one must have felt in summer the
heat to be at times almost insupportable, without any
apparent reason, as shown by the reading of the thermometer. This happens when the air is nearly calm and
moist; the air is already so moist that it cannot take off
our own moisture as we give it off in perspiration, and
so we say it is "sultry;" but only let the air get in
motion, even only by means of the Indian punkah or
huge fan swinging about, and then we feel cool, and
experience relief. Yet, the same hygrometric conditions
exist. It is only a very small amount of vapour and
heat that we force the air to take from us by the process.
But, should the air get *drier* with the same temperature, then evaporation from our skin takes place with
great activity, and we feel a marked sensation of cold;
and this result is as great a fallacy as the former. The
fact is, that, with the same temperature, and enjoying
an equal state of health, we experience, according to our
mere sensations, various changes of temperature, and so
our senses cannot guide us with regard to heat and humidity, as far as our own health is concerned, and much
less with respect to that of our plants. Therefore, the

hygrometer, properly used, and its indications attended to, may be made the means of preserving many valuable plants which might otherwise perish in an ill-regulated atmosphere.[1]

We presume that most of the superior gardeners are well acquainted with the nature of the proper climates of their plants. At any rate, such knowledge is necessary for their proper treatment, not only as to temperature, but also as to moisture. With this information, they can easily regulate the atmospheric condition of their conservatory. For example, suppose the temperature of the climate of the plants be 70°, and its mean state of humidity about 60 or 70 per cent. of the quantity of aqueous vapour which the air would contain if saturated, as before explained. Then, by consulting Mr. Glaisher's Tables, we find that, in order to secure this climate for the plants, the reading of the dry-bulb thermometer must be 70°, and the reading of the wet-bulb thermometer between 60° and 64°. In this manner we can match any climate under the sun, as Mr. Glaisher's Tables, which should always accompany every hygrometer, include every temperature and degree of humidity. We have only to look out for the temperature of the dry-bulb in the first column, and find, in the eleventh column, the degree of humidity required, and then, in a line with it, in the second column, we shall find the requisite reading of the wet-bulb to secure it.

Having made this discovery, all we have to do is to introduce a large surface of water, with a moveable cover, to regulate at pleasure the extent of evaporating surface, and then we shall have the means of obtain-

---

[1] Glaisher, *ubi supra*.

K

ing and securing continually the requisite degree of
humidity.

In like manner, when water is thrown on the walls
for the same purpose, the hygrometer will show the degree of humidity produced by the operation.[1]

The theory of aqueous vapour and of that of the
hygrometer are conquests of modern science, and they
open to every observer an endless field of the most interesting and useful research in nature.

[1] Glaisher, *ubi supra.*

CHAPTER X.

THE WONDERS OF EVAPORATION, AND ITS SIGNS IN PROGNOSTICATING THE WEATHER.

EVAPORATION, or the change of water into vapour, has the most important bearings on meteorology. In this respect it is one of the most effective of all the gigantic processes that are continually going on around us. Watery vapour is continually rising invisible in the air; meeting a colder stratum of the atmosphere, or the cold ridge of a mountain, it becomes condensed into mists or clouds; the fine particles of these unite into larger groups, and fall as rain, hail, or snow,—to be again evaporated by heat from the moist ground, or from rivers and seas. Even when otherwise invisible, its presence may be detected by its deposition as *dew*, and, probably, in the blue of the sky, and the gorgeous tints of sunrise and sunset, all the colours of which result from the vapours in the air in various conditions as to condensation or saturation,—for instance, the *red*, when it is on the point of condensation; the *yellow*, when it is already condensed into clouds; the *grey*, when the air is saturated with moisture. There is very little doubt that vapour is also intimately connected with the twinkling of the fixed stars. Atmospheric electricity is largely

K 2

due to evaporation directly as well as indirectly, on account of the amount of vapour contained in different currents of air. Of the immense energy of vapour some idea may be formed when we state that in the tropical regions, a fall of rain often takes place over a vast extent of surface, sufficient in quantity to cover it with a stratum of water more than an inch in depth. If such a fall of rain were to take place over the extent of a hundred square leagues, as sometimes happens, the vapour from which such a quantity of liquid would be produced by condensation would, at the temperature of only 50°, occupy a volume one hundred thousand times greater than that of the liquid; and, consequently, in the atmosphere over the surface of 100 square leagues it would fill a space of 9000 feet, or nearly two miles in height,— whilst the extent of the vacuum produced by its condensation would be a volume nearly equal to 200 cubic miles, or to the volume of a column whose base is a square mile, and whose height is 200 miles! Now if, of all the causes by which winds are produced, the most frequent is the sudden condensation of vapour suspended in the atmosphere, we can easily, after this explanation, comprehend the prodigious force of the tropical hurricanes resulting from such condensation.

In the temperate zone, the average annual evaporation is about 37 inches, but in the tropics from 90 to 100. Contrary to the received opinion, Kaemst maintains that the higher regions of the atmosphere are not drier than the lower. While for a period of nine weeks, the air at Zurich did not contain, on an average, more than 74·6 per cent. of vapour, it contained 84·3 on the Rigi mountain. Indeed, as will appear in the sequel, it is probable that at times there is an accumulation of vapour in the upper regions of the atmosphere, the existence of which can alone account for certain thunderstorms.

Such are a few of the curious and interesting facts connected with evaporation, and as all of them are intimately related to the weather, it is obvious that evaporation should be noted at all times by those who are interested in the weather, by means of such instruments as are at our command. For, although it is not likely that with a little water exposed in a vessel of a few inches diameter, we can obtain a complete solution of the problem as it is set before us on the great scale of nature,—still approximations by an evaporation-gauge will serve, in many instances, to repay us for the trouble, by warning us of effects before which both the barometer and even the hygrometer may give no sign. We will give a few instances :—

In general, a comparative dryness, as indicated by the hygrometer, is connected with evaporation and fair weather, and an indication of moisture with rain,—regard being had, in both cases, to the mean of the season. But there are exceptions to be noticed. In summer, when the precipitation of moisture is actively going on above, and thunder-clouds are already formed, the air below may continue—from the intense heat and the arid state of the soil—hygrometrically dry. A morning may be cloudy, then fine, the hygrometer showing the mean humidity of the period,—in the evening, heavy rain, with hail, thunder, and lightning, the hygrometer indicating before the storm a considerable *decrease* of moisture. In winter, on the contrary, the air is sometimes very moist for a considerable time, without rain, chiefly during the prevalence of foggy days and frosty nights, with a high barometer. Such an instance occurred for the space of more than two weeks, from the 21st of December to the 8th of January, in the year 1818.

In order to be always on our guard against such indications, and to enable us still more accurately to predict coming weather—especially the approach of thunderstorms—we may have recourse to an *Evaporation Gauge*, which should be the companion to every hygrometer.

There are few *days* in the whole year in which some vapour is not given off from the gauge, but the process is apparently suspended whilst *dew* falls at night. On the other hand, a state of the air analogous to this appears to be the cause of its complete interruption by day, chiefly about the period of the year before mentioned, namely, the last week in December and the first week of January.

Evaporation is not always suspended during *rain;* the rate is, however, much less on those days in which rain falls, and it is liable to a rapid increase immediately afterwards.

Sometimes an excess in the rate is found to *precede* rain, whether from the agitation of the air, or the effects of electricity, or from both causes combined.

The calm which attends a change of wind sensibly lowers the rate; which also decreases, as might be expected, upon the cessation of a wind which has been blowing steadily for some days.

A moist current of air flowing in upon us will sometimes check the evaporation, although no rain be produced from it.

Of course there is a gradual increase in the rate of evaporation with the elevation of daily temperature, and a decrease in the contrary; and it is subject to variations in windy weather.

These prominent facts alone are sufficient to show

that an evaporation gauge is capable of being made a reliable weather-instrument; but there are many other, equally important, meteorological data supplied by it, tending to show that we may be able to draw up a few rules to interpret its indications.

The *smallest* monthly results of evaporation furnished by the instrument were found at the approach and during the continuance of the great frost of 1813-14; and there was a striking example of the retarding effect of a moist air on the process of evaporation. In the last month of 1813, with great *fogs* prevailing, the mean temperature being 38° 43', the evaporation was 0·21 in. (in all probability the lowest amount in ten years); but in the first three months of 1814, the frost having set in with rigour, and cleared the air, there was a gradation of increasing results thus—0·25, 0·36, 0·83 in., with the mean temperatures—26° 71', 33° 17', and 37° 82', all inferior to the former, and the first of them almost 12° below it. This difference in the effect was plainly due to the extreme dryness of the currents prevailing in the latter period. Indeed, the most intense *cold* is insufficient in itself to put a stop to the formation of vapour. Ice evaporates freely during *a clear frosty night*, as Luke Howard found by direct experiment repeatedly. And the rate of such evaporation is rather astonishing for the circumstances. In one of his experiments, a circular area five inches in diameter, lost 150 grains between sunset and sunrise. This is at the rate of more than 8000 troy pounds of ice, or nearly 1000 gallons of water, from an acre of surface, in that time. The absorption of heat necessary to the composition of so much vapour within a small space of atmosphere, must be prodigious.

With these facts before us, we need not wonder to hear that a moderate fall of snow is sometimes entirely taken up again. during a succeeding northerly gale, without the least sign of liquefaction on the surface. As must be familiar to all observers, the surface in deeper snows after awhile becomes curiously grooved, scooped, and channelled from the same cause. The effect is most conspicuous around the trunks of trees and near the interstices of palings,—in short, wherever the stream of air acquires force in a particular direction. A little observation will satisfy any one that the snow is not removed on these occasions merely by being driven before the wind.

In accordance with these facts, a sensible change in the air to *a dry state, after damp, foggy weather in winter, may be always placed among the indications of approaching frost.*

As before stated, an evaporation-gauge will furnish us with the evident sign of an approaching thunderstorm, namely, by a rapid *decrease* of the daily evaporation during hot weather.

The greatest evaporation noted by Luke Howard in one day (with a single exception) occurred on the 17th of May, 1809. On that day, the amount was 0·39; on the following, 0·28; on the next, 0·14 inch; the corresponding mean temperature being 67°, 70·5°, and 64°, and consequently furnishing, in respect of heat, no adequate cause for the decrease. But in the evening of the 19th, occurred one of the most tremendous thunderstorms on record (May 19th, 1809).

One is led to believe with Luke Howard that on this occasion, the local influence of heat, aided by an electric charge in the air, had suddenly raised, as it were, a

mound of vapour into the elevated regions which it
rarely visits in these latitudes, and where it is subject—
from the contiguity of an intensely cold medium—to
extensive and complete decomposition, in which seems
to lie the true cause of the prodigious development of
electricity manifested on those occasions.

It was therefore gaseous pressure which checked the
evaporation and caused the rapid *decrease,* thus indicative of the approaching thunderstorm.

A similar indication occurred in the following June of
the same year, 1809, when the decrease in three days
was as follows :—0·33, 0·26, 0·19, followed by a tempest
of wind and a week's wet weather.   But in this case,
unlike the former, the barometer, temperature, and sky
furnished concurrent indications.

As the temperature advances in the fore part of the
year, the evaporation on the whole increases steadily ;
but in particular years, it receives a check in some part
of the spring, which is afterwards made up by a sudden
increase.   The cause of this is sometimes obvious in the
variations of temperature.   The rate is likewise occasionally kept down in this part of the year, as in the latter
months, by frosty weather.   The very great increase in
a fine spring may be possibly due, in part, to the electric state of the air in such seasons ; for, although electricity, in the low degree in which it is applied by nature
at the earth's surface, may not sensibly promote the actual emission of vapour from water, it may tend greatly
to increase the retentive power of the air by rendering
the particles of the mixture of gases and water in a
higher degree mutually repulsive, or, in other words, by
keeping up the *elasticity* of the atmosphere.

Not only is the low rate of evaporation in spring due

to the state of the temperature, but we may say that
spring is actually made colder *because of the evaporation*, small as it is. It cannot be doubted that the
sharpness of our N.E. breezes in spring is, in a measure,
the result of their excessive dryness, relatively to the
temperature that prevails,—in consequence of which
they abstract the heat from the animal system by means
of the moisture on the skin, which they convert, with
peculiar rapidity, into vapour, the result being the well-
known chilly sensation.

In the latter part of spring, the evaporation-gauge
sometimes indicates an abundant supply of vapour,
when, in fact, very little is poured into the atmosphere
from the earth, the surface, and even some considerable
depth under it, being already dried by the sun and wind.
It is then that we perceive the effects of *natural irrigation*, carried on by means of the vapour diffused in the
daytime from canals, rivers, etc., and condensed by
night in copious dews, which descend on the neighbouring herbage.

Should the season afterwards prove showery, a great
quantity of the first water that falls is vaporized by the
heated earth, with a rapidity of which, again, the gauge
gives no adequate indication. This vapour may even
continue to be thrown up after the air has begun to approach towards saturation, and thus contribute to the
formation of the *next* rain; and the water may be thus
driven from the earth to the clouds, and returned again
in rain, until the surface, being cooled down, is prepared
for drying under the solar rays by a drier current.

The sudden change from a dry to an extremely moist
state of the air, immediately after our spring and summer showers, is often sufficiently obvious to be detected

by the most superficial observer. It is generally due to this sudden and copious production of vapour at the surface. The spring and summer are our most variable seasons in point of dryness, as indicated by the hygrometer and evaporation-gauge.

In the autumn, or rather at the approach of winter, the rate of evaporation declines with great rapidity. The commencement of a saturated state of the air, while as yet precipitation has not generally commenced, gives to our autumnal weather that delicious *softness* which is the reverse and the compensation of those keen blasts which so often attend the vernal season.

This state, however, does not continue long. On the approach of the *first frost*—indeed, during a great part of our ordinary winters—the earth and waters retaining a temperature somewhat *above that of the air*, continue by the force of this inherent warmth to emit vapour. This is continually undergoing decomposition, and it fills the air with a *mist*, which, when by no means dense enough to constitute what we call *fog*, would yet appear to an observer stationed above its limits as a white veil thrown over the whole face of the country,—thicker, indeed, in the valleys and along the course of the rivers, but nowhere in the London district surmounted by the land.

The greatest evaporation in a month, by an elevated gauge, recorded by Luke Howard, was about six inches. In this case, a number of favourable circumstances appeared to concur, namely, a high spring temperature, succeeding to protracted cold ; dry winds ; and an abundant electricity.

The smallest monthly results were found at the *approach and during the continuance* of the great frost of 1813–14, as before given.

The cessation of the excessive heat in July, 1808, was attended with the following rapid decrease in the daily rate of evaporation at Plaistow, although the reaction in the atmosphere took place about Gloucester; the hottest day giving evaporation 0·35, the four following were 0·31, 0·27, 0·20, 0·16 inch; and this without more than a few drops of rain in the immediate vicinity of London.

The greater the evaporation in any month, the less the rain, and *vice versâ*.   Thus, in December, 1809, the evaporation was 12·1 inches, and the rain was 0·12 inch.

In December, 1814, the evaporation was 0·68 inch, and the rain was 3·35 inches.

The following table shows, according to Luke Howard, the average evaporation and mean temperatures of the seasons :—

	Evaporation.	Mean Temperature.
Winter	3·587	37·20
Spring	8·856	48·06
Summer	11·580	60·80
Autumn	6·444	49·13

Evaporation is not in strict proportion to temperature. In spring it is *augmented* by about a sixth part; and in summer about one fourth part, in consequence of the *dryness* of the air in these seasons; while in the three months of autumn it is *lessened* by more than a sixth; and in those of winter by considerably more than a third, owing to the *dampness* of the air.

RULES FOR THE EVAPORATION-GAUGE.

1. A great decrease in the rate during hot weather prognosticates a thunderstorm.

2. A decrease precedes, in general, all stormy weather, with or without rain.

3. *Sometimes* an excess in the rate precedes rain.

4. An increase of evaporation during rain prognosticates fine weather, although the hygrometer may show considerable humidity.

5. A low rate of evaporation indicates continuous rainy weather at hand.

6. A dry state of the air, after damp, foggy weather in winter, may be always safely considered an indication of approaching frost.

7. The rate of evaporation is lowered in the calm which precedes a change of wind.

8. Evaporation decreases when a wind, which has blown for some days, is about to cease.

9. If a moist current of air be blowing, the *hygrometer* will probably indicate a critical degree of humidity, but if the *evaporation* is checked there will be no rain.

10. In the latter part of spring the gauge sometimes indicates an abundant supply of vapour, when, in fact, very little is poured into the atmosphere from the earth ; the surface, and even some considerable depth under it, being already dried by the sun and wind.

11. On the other hand, as the thermometer and hygrometer simply indicate the condition of the air at the place in which they are situated, we may have a clouded sky with a *dry* hygrometer, if *the decrease of vapour is more rapid than usual.* If the reverse takes place, and the air is unusually hot above, the hygrometer may be moist with a clear sky, when no rain is likely to fall. The rate of *evaporation* can alone guide us to a right conclusion in these circumstances, not unfrequent in summer, and the latter part of spring. At any rate, the state of the sky should always be considered when we interpret the indications of the hygrometer. The rate of evaporation must be always taken into account.

As it is very desirable that we should have an instrument capable of furnishing the *rate of evaporation by inspection,* we have ventured to propose the following.

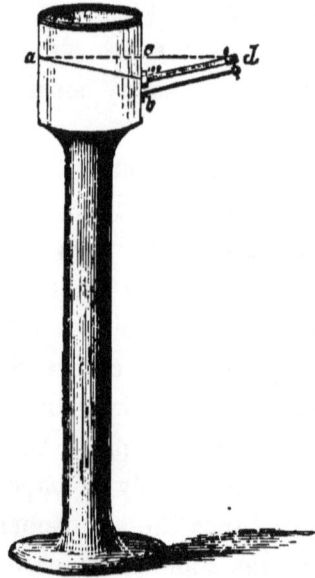

Steinmetz's Vaporimeter.[1]

This instrument is founded on the principle called the *hydrostatic paradox,* which, however, is no paradox at all, because the action of a fluid is downward, not upward, and so any quantity of fluid, however small, may be said to "counterpoise" any quantity, however large. The principle is seen in the kettle or the teapot, in the spouts of which the water or the tea is always on a level with the water or tea in the respective vessels. Just as the level of the water in the vessel is lowered, so is that of the portion in the spout.

[1] This instrument is made by Casella, Hatton Garden.

It will be seen in the figure that the diagonal *a b* of the inch of water *b c* is the extended measure of the said inch; that is, the inch of depth is extended over the space of the diagonal, which varies in length according to the depth of the fluid mass.

This diagonal, therefore, gives us, in the case of a five-inch diameter in the evaporating vessel, rather over 5 inches for extended graduation, which, being divided into *hundredths*, will accurately show the evaporation of the water in the vessel. The tube *d* is exactly this diagonal thus graduated; and the amount of evaporation may, at any time, be read off by inspection. The amount of evaporation from the tube itself may be disregarded, as it is checked by the brass-cap enclosing the mouth. This can never amount to as much water as is lost by the ordinary measuring process in ascertaining evaporation.

The extent of evaporation in a day rarely exceeds one inch; and so this extent of graduation will be sufficient, the instrument being set, from time to time, as the water level is lowered.

The Vaporimeter should be shaded or isolated by padding, in the height of summer, exposed to sunshine and the winds. The correction for rainfall into it must, of course, be made by means of the rain-gauge. Thus, the important fact of the extent of evaporation during rain may be always ascertained.

In windy weather the agitation of the water in the tube will be stopped by merely placing a piece of board or slate on the top of the vessel, until the observation is made, which requires but an instant. Of course the instrument must be levelled.

As it is in open weather that the chief meteorological

inferences from the rate of evaporation are to be made, the instrument should be removed during continued frost. A casual night-frost will not materially interfere with the action. The investigation of vapour-formation is, however, mainly a summer and autumnal occupation.

Of course we may obtain the evaporation by measuring off the water exposed in a vessel of the diameter of ordinary rain-gauges, say five inches diameter, when the amount evaporated will be found by the rain-gauge measure. Only the amount of trouble in measuring half a gallon of water to find one hundredth of an inch of evaporation is rather prodigious,—to say nothing of the loss by the actual measurement.

It should always be remembered that a great and sudden evaporation often precedes rain, as does also a sudden suspension of it—when the whole air becomes moist. The following table gives :—

THE AVERAGE EVAPORATION AND RAINFALL OF THE MONTHS IN LONDON.

Month	Evaporation		Rainfall	
January.	0·83 in.		1·95 in.	
February	1·64 ,,	,,	1·48 ,,
March	2·23 ,,	,,	1·29 ,,
April	2·72 ,,	,,	1·69 ,,
May	3·89 ,,	,,	1·82 ,,
June	3·50 ,,	,,	1·92 ,,
July	4·11 ,,	,,	2·63 ,,
August	3·96 ,,	,,	2·12 ,,
September	3·06 ,,	,,	1·92 ,,
October	2·20 ,,	,,	2·52 ,,
November	1·16 ,,	,,	2·99 ,,
December	1·12 ,,	,,	2·43 ,,

The total of the year is 30·50 inches. Invariably, the greater the *evaporation* the less the *rain,* and *vice versá,* in every month and on all occasions.

The above is Luke Howard's estimate.   The following is Daniell's:—

	Evaporation	Rainfall
January.	0·62 in.	1·56 in.
February	0·84 „	1·45 „
April	2·52 „	1·55 „
May	3·99 „	1·67 „
June	3·75 „	1·98 „
July	4·24 „	2·44 „
August	4·03 „	2·37 „
September	3·03 „	2·97 „
October	2·13 „	2·46 „
November	1·17 „	2·58 „
December	0·62 „	1·65 „

The great differences in these results from those of
Luke Howard, above given, may be explained by the
fact that Daniell's estimates are based on observations
made in the garden of the Horticultural Society at
Chiswick; and we apprehend that Luke Howard's averages will be found more generally to correspond with
observations in London; at any rate, such is the result
of our experience.

In recommending the use of an evaporation-gauge,
we are perfectly aware of the imperfection of any such
instruments, although we do not endorse Daniell's sweeping objection to them to the exaltation of the theoretical
calculations deducing the evaporation from the indications of the hygrometer.   No doubt the conditions which
modify evaporation vary *ad infinitum,*—on the land and
on the water,—in sunshine and in shade,—as land is
more or less clothed with vegetation, or as water is
more or less deep; but if, as Daniell says, "the evaporating gauge, so far from representing the circumstance
of these bodies which yield the great body of vapour on
the earth's surface, probably does not correspond, in all

L

essential particulars, with a dozen puddles in the course of a year,"—we do not see why the same line of argument should not, in a measure, equally apply to the rain-gauge, in the results of which, as is well known, important differences occur in different localities. As an indication of weather, we contend that the *rate* of evaporation is most important, and that such information, by direct means, however imperfect, is valuable to the meteorologist.

Oddly enough, the greatest differences in the estimates are in the *rainfall*, not in the *evaporation*; and it is quite certain that Luke Howard's means of ascertaining the latter was that which Daniell deemed only corresponding "with a dozen puddles in the course of a year." At any rate, we find that our Vaporimeter's esults tally with those of the hygrometer by calculation, as previously explained (p. 60). Calculation will do, no doubt, to discover evaporation; but, as Daniell himself observes elsewhere—in objecting to the *Wet Bulb* hygrometer—"Little is gained, or more frequently everything is lost, by this transfer of labour from the observer to the computator."

Finally, the *Vaporimeter* is an accurate gauge of *Dew-fall*, and the precipitation of water during *fog*—as we found it to be during the month of December, 1866 —showing the amount by inspection.

CHAPTER XI.

THE TEMPERATURE OF THE SOIL.

In the previous sections we have drawn attention to a
variety of signs, whereby the nature of coming seasons
may be foreseen with more or less probability; and it is
evident that we lay the greatest stress on the necessity
for studying and observing the winds; for, from whatever cause they may proceed, all of them directly influence vegetation for good or for evil, because *every
wind has its weather.*

In a thing of such variety as the wind it may seem a
hopeless task to arrive at anything like definite conclusions; but even the variety of the winds admits of an
easy and comprehensive explanation, and one that may
serve to clear up not a few doubts and difficulties regarding their play in different localities.

The wind always moves in the direction of what is
called " a great circle;" that is, one which, were it continued, would divide the globe into two equal parts; so
that, if the wind proceeded from two of the cardinal
points of the compass—that is, either from north or
south—it would then retain its name over a great part
of the globe. In that case it would blow along the me-

ridian or line passing from pole to pole. But should the
wind proceed in any other direction, it will seem very
different in whatever distant places it may pass over, unless under the equator when it blows due east or west.

The reason is, because, according to the principles
of geography, all the lines (called rhumb-lines) which
give names to the winds, between the equinoctial and
the poles, are not straight, but spiral lines, like a
corkscrew.

For instance, should a wind set out from the equator
in the direction of an angle of forty-five degrees, then,
in order to retain the same name—south-west, for example—it ought to cross the meridian of every place it
passes over, in the same angle. But if it keeps right
forward, this is impossible; because the meridians are
not parallel, but inclined lines—all meeting in a point
at either pole. Therefore, the wind which is termed
south-west in our latitude will always carry a different
name in another. Such is the explanation of the variety
of the winds in general, in addition to the numerous
local causes which may give rise to them in all parts of
a country. It is therefore obvious that it is in our
power to trace, so to speak, the story of any wind,
either from or back to its origin; and no doubt this
must be done if we ever set up a true theory of the
winds. But, fortunately, as before alluded to (and it
cannot be too much insisted on), we have a safe guide
to the true nature of every wind by means of the hygrometer. With this instrument we shall always be able
to prove the true identity of every wind, in spite of its
various *aliases*.

But the discovery of this important fact is not all that
must be secured for the purposes of agriculture. The

winds are constantly fanning the soil, and if they do not warm or keep it warm, they carry away its warmth; and so the relative prevalence of certain winds in various districts must have a great deal to do with their vegetation. The heat of the soil must either be inherent or acquired from sunshine; and every wind either leaves the soil warmer or colder,—not directly, but either by *convection* or by not being in condition to *abstract heat.*

Now, whether such is the case must be discovered by direct experiment in the soil itself; and, considering the matter in all its bearings, it is surprising that so little attention has been paid by farmers and gardeners to so important a matter as the temperature of the soil below the surface.

The temperature of the air is admitted to be an important consideration; but the temperature of the ground, which is certainly not less important as regards the health of plants, is scarcely recognised as a necessary point of investigation. We seem satisfied to look at the outer garment without giving a thought to the under-clothing, upon which, for the most part, robust health mainly depends.

Latent heat, as it is learnedly called, or that inherent heat not perceptible to sensation, which exists in things positively cold to the touch, plays a most important part in the economy of vegetation; indeed, many plants can better stand against a cold season than resist the effects of a damp or cold soil; and there certainly is not one that would not have a better chance, during a bad season, for having a warm soil at its roots—warm under-clothing next to its skin. Take the hop plant, for instance. How many more growths have been saved, and how many poor growths improved, since the ground has been paid more attention to by draining, and manuring with a bad conductor of heat,—namely, pieces of old *woollen rags*, which keep in the heat or prevent it from escaping, just as the material does in our own under-clothing! This shows what might be done for other kinds of cultivation.

Moreover, it seems very probable that the various dis-eases, and blights, to which all plants are more or less subject, proceed from the *deficient warmth about their roots*, so that when the parching easterly and north-easterly winds come upon them, they are not provided with the means of resisting the cold, their vessels shrink, their fluids become vitiated, their blood, so to speak, is corrupted, and inevitable disease is the consequence.

To discover the temperature of the ground below the surface, ground-thermometers are employed. We should like to see such instruments multiplied and planted in every establishment devoted to agricultural pursuits throughout the land. There is not a farm where they would not be highly useful, as they would indicate the precise temperature of the earth at any desired depth below the surface at which the bulb of the thermometer is placed; and thus the farmer would have a guide in *draining his land*, or for regulating the temperature by the addition of a warmer or colder bed, according to the requirements of the particular crop which it may be in-tended to put in. It seems that the idea of such instru-ments being too "scientific" for the practical man, has prevented their use; but this is quite erroneous, as there is really no more scientific difficulty in using this instru-ment than in the thermometer or barometer, as we shall endeavour to show by a few remarks on the instrument and the mode of using it.

Of the amount of heat which the earth annually receives from the sun, some portion is lost by radiation into space, or is taken off by the air or wind in contact with the soil. Some is absorbed by the soil and gradually descends, being conducted by the earth to a considerable depth below the surface. But the heat thus absorbed is slowly given off again during the *spring* months, to be restored by the sun when the hot season comes round again. The amount of heat thus received and garnered up for the benefit of vegetation, will depend upon the nature of the soil. Those soils which take it easily and rapidly will part with it as easily and rapidly, and it is only by means of the ground-thermometer that we can discover this very important quality of the soil, although, of course, in theory a geologist will tell you at once the capability of your soil as to the receiving and retention of heat. And, for our part, we should never take a farm without having a thorough geological knowledge of its acres, any more than we should buy a horse without knowing its points and defects. As in most things here below, the outside or surface is apt to deceive us; and it is only by going deeper into the soil that we get to know beforehand whether we are likely to succeed with it in any cultivation we may have in view. In the varying seasons, however, the best of lands may prove deficient in warmth, and so we had better have recourse constantly to the ground-thermometer, as it were, to feel its pulse like the doctors, in order to see what is to be done for it.

The thermometers for agricultural purposes need not be of that degree of delicacy, nor be sunk to so great a depth as those used for philosophical experiments. Generally it will be found that about two or three feet will be deep enough; but of course this must be determined by the kind of crop for which the ground is prepared. Some of your roots grow to a prodigious depth, although very tiny, and we apprehend that much depends upon securing an adequate amount of warmth to *all* the roots of plants. At all events, look to the temperature of the soil at a sufficient depth to keep all warm below.

If the bulb of an ordinary thermometer, suspended in the air, be coated over with a spherical or round covering of some kind, and the index of such a thermometer be observed, we shall find that it is far less affected by changes of temperature than one whose bulb is freely exposed; and the daily range or difference between the highest and lowest readings during the day will grow less and less as we increase the covering in thickness, but particularly so if the covering is made of some substance that is a bad conductor of heat.

Now, by sinking our thermometer into the bosom of the earth, we arrive at a similar result: the earth will act as a shield, protecting the thermometer from the influence of the great and sudden changes to which the *air* is liable, and causing it to be affected by the heat conducted to its bulb by the earth only. Instruments so placed will then exhibit the exact amount of the solar heat absorbed by the earth at the depth at which the bulb is planted. Care must be taken to bed it up well, or provide it with a shield to prevent rain-water from swamping the bulb; if this happens, the indication will be false, of course.

The temperature at the surface of our globe is, from a variety of causes such as we have mentioned, liable to great and sudden fluctuations; but this is not the case as we get below the surface. The lower we descend the smaller the fluctuations become, until at length we get to the stratum or layer of *invariable* temperature, which varies in different soils from 50 to 100 feet below the surface.

We have before us a table showing the mean monthly temperature at various depths, and in the air, at Greenwich, from an average of thirteen years; and it appears from this table that the full effect of the solar heat received by the earth in the summer months does not reach to the depth of twenty-four feet until about the end of November or beginning of December; and that the lowest temperature is not attained till June, by which time the absorbed heat has been radiated, and the ground receives a fresh supply. The following are the dates of the middle of the warm and cold periods for six thermometers, from an average of thirteen years, at the Royal Observatory, Greenwich :—

Thermometers.	Middle of warm period.	Middle of cold period.
(1) Air	July 21	January 20
(2) 1 inch underground . . .	„ 26	„ 24
(3) 3 feet „ . . .	August 9 . . .	February 8
(4) 6 „ „ . . .	„ 25 . . .	„ 24
(5) 12 „ „ . . .	September 25 . .	March 27
(6) 24 „ „ . . .	November 30 . .	June 1

From this table, we see that the thermometer whose bulb is only covered by one inch of soil reaches its highest and lowest readings at dates differing only a day or two from the lines of highest and lowest, as marked by the thermometer free in the air. The three-feet thermometer is about nineteen or twenty days behind the instrument in the air; that six feet below the surface is retarded by still more; and this goes on increasing with the depth, until at twenty-four feet the times of maxima and minima are retarded between four and five months.

The range of the mean monthly temperature at the different depths is as follows :—

Thermometer in the air	29·8	degrees.
„ sunk 1 inch	25·4	„
„ „ 3 feet	21·7	„
„ „ 6 „	15·4	„
„ „ 12 „	9·5	„
„ „ 24 „	3·4	„

As before observed, the geological peculiarities of the soil in which the thermometers are sunk will make some difference in their readings, some strata or layers of the soil having a much greater power of conducting heat than others. It takes about five or six days for the impression of heat to penetrate through one foot of earth, subject to slight variations in different soils.

From these facts, the reader will perceive how much warmer the roots of plants are during the winter than the portions that remain above the ground; and the contrast becomes more striking when the winter is very cold. When we are hardly able, with the greatest care, to keep life in some of the plants above, those beneath are luxuriating in a temperature many degrees warmer, provided they have fair-play, and are not over-supplied with moisture, the excess of which makes lands " cold" and ungenial to vegetation. Thus we can understand why the natives of the ice-bound regions of the north retire, when their short summer is over, to homes below the surface of the frozen ground to pass the long winter.[1]

The practical application of these facts and principles

---

[1] Chiswick, ' Agriculturist's Weather Guide.'

should be easy enough. Having by means of your ground-thermometer discovered the maximum or middle of the warm period of your land, the further it is from the maximum or middle of the warm period in the air, the greater the certainty that your land is in good condition of heat, and that the cold period will be delayed, so as to counteract the effects of a cold, untoward spring.

The effect of judicious drainage in thus getting the land into good heat—that is to say, in point of fact, increasing its capacity for absorbing heat—has been generally admitted; but we are tardy as yet in seeing the immense advantage, in the same line, that is likely to result from *deeper subsoil cultivation.* If anything will tend effectually to counteract the stress of untoward seasons, it is, we believe, deeper cultivation, of which nature gives us a hint in the length to which the tiniest roots will delve below the surface, seeking better conditions of life and growth. In proof of this, we have only to refer to cultivation in Japan, one of the wettest of countries, but with an *arable* soil of some twenty feet deep, instead of only six or eight inches as in England, and consequently incapable of suffering either from drought or excessive rainfall, and producing not only enough for her population—at least equal to ours—but also for exportation; and more than that, for *any* cultivation, at will, even *rice*, which requires flooding.

CHAPTER XII.

THE EFFECT OF DRAINAGE ON THE TEMPERATURE OF SOILS, THE CLIMATE, AND RAINFALL.

" BEING on a journey," says Luke Howard, the father of British Meteorology, " I made some observations which seem to prove that showers are at times determined in their extent by the nature of the soil and surface. At London, on the evening of the 20th, we had a small rain without thunder, notwithstanding the previous heats. The roads were moist on the 21st, when we set out, the soil being loam. Between Barnet and St. Alban's on the chalk, they were dusty, with only the marks of large drops on them. Some miles beyond the latter place, we found the road, on sandy loam, watered by a shower going before us. The further boundary of this shower was Dunstable. On the calcareous downs, leaving Dunstable, no rain appeared to have fallen, the dust flying freely; but at the foot of the hill, where the new road is cut, we found the soil (a sandy loam) again well watered, precisely to the boundary between it and the chalk. Arriving at Woburn, we learned that showers had fallen there, on the sand, in the course of the afternoon, attended with hail. Thus the rain of the day had been clearly determined to the sand and clay, leaving the chalk dry."

This is a very curious observation, and possibly it may have some relation to a general law throughout the mineral kingdom, by which each kind of "rock" or stratum attracts or repels moisture. But what degree of truth there may be in the idea must be left to be determined by Mr. Symons and his thousand rain-gauges widely spread over the country. Meanwhile the rainfall bids fair once more to excite agricultural interest by its excess, as during the past few years it has more generally done by its deficiency, and the relation of the soil to temperature in lands that are water-logged and in those that are drained, becomes a subject of first-rate agricultural importance.

Some four years ago, the Marquis of Tweeddale offered a premium of £80 for the best sets of observations on the difference of the temperature of drained and undrained land, at different depths below the surface, compared with the temperature of the air, at four feet above the surface, in the same locality. The object of the observations was to inquire whether drainage raises the temperature of the soil, and improves the climate, and if so, to what extent the produce of the land has thereby been increased.

The conditions of the experiments were calculated thoroughly to test the question. Competitors for the premium were instructed to see that the soil of the drained and undrained land was as nearly as possible the same in nature and quality; and it was stipulated, in order to secure results comparable with each other, that the cultivated fields experimented on should be under the same crop, and have received the same tillage.

On mountain pastures, the thermometers were to be placed at the depth of 10 inches, and in cultivated land 10, 20, and 30 inches below the surface. These thermometers were to be respectively 16, 26, and 36 inches long, projecting 6 inches above the surface; so that the bulbs were of the required depths, care being taken to keep the bulbs clear of the drains, so as to give the true temperature of the soil affected by the drain. The thermometers for indicating the temperature of the air were to be placed, as before stated, 4 feet above the surface of the ground, and protected from sunshine and radiation in such a manner as to allow at the same time of free ventilation. The instruments were to be observed and registered daily at nine A.M. All who may wish to institute similar inquiries cannot do better than follow this method.

We will now proceed to state the leading results of this important investigation :—

Effect of Drainage on the Temperature of Arable Land.—The following observations were made on a light sandy soil under a rye-grass crop, in a field sloping to the west, about 1 foot in 40, and about half a mile distant from the east shore of Loch Fyne. The drains were 2½ feet deep.

First, at the Depth of Ten Inches.—It must be remembered that changes of temperature are much more marked at this than at greater depths.

Effect of a Sudden Fall in the Temperature of the Air.—On the 20th of October the temperature of the drained land was 46·8°, and of the undrained 46·2°; the difference being only 0·6°. But on the following night the air fell to 27°, being 11° lower than on the previous night; and next morning the temperature of the drained land fell to 44·8°, and the undrained to 42·2°; the drained land being thus 2·6° warmer than the undrained.

In summer, on the 29th of June, the temperature of the drained land was 61·5°, and of the undrained land 59·9°. On the following night the air fell to 48·0°, being 6·0° lower than on the previous night; and the next morning the temperature of the drained land fell to 59·5°, and the undrained to 56·9°. Thus a fall of 6·0° in the temperature of the air was followed by a fall of 2·0° in drained land, and 3·0° in the undrained land.

Effect of a Continued Fall in the Temperature of the Air.—During the time the observations were carried on there occurred eight well-marked depressions in the temperature of the air; and in every case but one the temperature of the undrained land fell to a greater extent than that of the drained land, but even in this instance the drained land was still warmer than the undrained by 0·3°.

In a *Protracted Frost* of three weeks the want of drainage reduced the temperature of the soil, at the depth of 10 inches, nearly half a degree.

In a *Sudden Rise of the Temperature of the Air* the drained land rose 1·0°, and the undrained 1·5°, but then, before the warmth set in, the undrained land was 2·4° colder than the drained. Moreover, on the following day the air rose to 72°, and next morning the drained land had risen to 61·5°, while the undrained land had fallen to 58·4°, being now 3·1° colder than the drained land. No rain fell on these days.

In a *Continued Rise of Temperature of the Air*— leaving out the cases where a copious rainfall raised the temperature of the undrained land—the mean temperature of the drained land was invariably higher than that of the undrained land. The mean excess was on one occasion 1·4°, and on three others about 0·7°, which are considerable amounts when spread over six days, as in the observation. In long-continued dry, warm weather the drained land was often 2°, sometimes 2·5°, or even 3·0° (though rarely) above the undrained land.

As may be expected, the drained land is not so much affected by rain as the undrained, especially in winter. On almost all occasions when rain falls during this season its temperature is higher than that of the land. In the case of the drained land, the drains quickly convey it away; but in undrained land it remains in the soil, and the temperature of the undrained land is by this means raised above that of the drained land. In one instance, within twenty-four hours after 1·35 inch of rain fell, the undrained land rose to 44·9°, whilst the drained land reached only 44·0°.

There are, of course, warm and cold rains, and their effect tells differently on drained and undrained soils. On the 27th of July 1·75 inch of rain fell, and the next morning the drained land had fallen 0·8°, the undrained 1·0°. This rain was probably colder than the soil, and thus we see the advantage of draining. Three days after 0·63 inch of rain fell, and next morning the drained land had risen 0·3°, and the undrained land 1·3°. This was probably a warm rain, and the undrained land had the advantage, but it signifies very little, when we remember the prevalence of cold rains during the early growing season. On the other hand, these different effects might have been caused by the time the rain fell, for if it fell during the day, when the surface of the soil is warm, it would carry part of the heat with it down into the soil; and if it fell during the night, when the surface is cold, an opposite result would follow. The effect of sleet, however, showed the superior condition of the drained land, the effect of sleet showers having lowered the temperature of the drained land only 2·0°, but that of the undrained land 4·0°. Of course, by remaining longer in the undrained land they cooled it more.

During a thaw the temperature of the drained land continued higher than that of the undrained—for the reason above stated; and it was only after a copious rainfall that the undrained land showed a higher temperature—doubtless owing to the warmth of the rain, as before suggested.

At Depths of Twenty and Thirty Inches.—At these depths the great advantage of draining becomes strikingly evident. During the prevalence of frosty weather in February for three weeks, the mean temperature of the drained land at 20 inches was 35·2°, and the undrained 34·7°; and at 30 inches, 36·0° and 34·7°. Hence the effect of drainage at the lower depth was to keep the soil 1·3° warmer during a protracted frost—an immense benefit to vegetation. Moreover, whilst at 10 inches the rain raised the temperature of the undrained more than that of the drained land, it is just the reverse at 20 inches, at which the drained land had the advantage. It was the same with snow, the effect of which was to prove that the cold resulting from a covering of snow is more quickly transmitted through undrained than through drained land.

The mean annual temperature at each of the three depths was greater in the drained than in the undrained land, the amount being, at 10 inches, 0·8°; at 20 inches, 0·4°; and at 30 inches, 0·7°. The annual mean tempe-

M

rature at nine A.M. is greater at 20 inches than at 10 or
30 inches; from which we may infer that the daily
wave of maximum temperature has at nine in the morning passed 10 inches, and penetrated to about 20 inches,
but has not yet arrived at a depth of 30 inches. Hourly
observations of the underground thermometers, conducted from time to time under different states of the
weather, would be of great service in determining the
march of the daily wave of temperature through different soils. Since changes of temperature are more
readily transmitted through undrained than through
drained land, it follows, from the small excess of the
drained land at 20 inches, that at nine A.M. the daily
maximum had penetrated to about 20 inches in the undrained land, but that in the drained land it was still
some distance from that depth.

The difference between the temperature of the drained
and undrained land varied greatly in different months
and at different depths. At 10 inches the excess is
much greater during the warm months, from May to
October, being 1·1°, as compared with 0·4° during the
other six months.   But at 20 and 30 inches the greatest
excess occurs during the cold half of the year, from October to March, being, at 30 inches, 0·9°, as compared
with 0·4° during the summer months; and at 20 inches
0·6°, as compared with 0·2°. It may be remarked, that
when the excess is greatest at 10 inches and least at 30
inches, the heat is increasing from the surface downwards; but when the excess is least at 10 inches and
greatest at 30 inches, the upper layers being coldest,
heat is passing from below upwards.

The temperature at 10 inches reached the maximum
on the 25th of July, and fell to the minimum about the

21st of February; and the mean temperature occurred about the 21st of March and 21st of October. At 30 inches those periods fell from one to two days later; except in April, when it occurred four days later, owing to the rapidity with which the temperature rose to the mean in that month, longer time being thus required to raise the temperature of the lower depths.

The coldest month was January, 34·2°; and the warmest July, 58·8°. In the drained land, the coldest winter and the warmest summer temperatures at the different depths were respectively 34·0° and 59·9° 35·4° and 59·1°; and 36·0° and 57·4°; and in the un-drained land, 33·9° and 58·9°; 35·1° and 59·2°; and 35·0° and 57·0°.

Thus, the difference between the mean temperature of the coldest and warmest months was, in the air, 24·6°; in the drained land, 25·9°, 23·7°, and 21·4°; and in the undrained land, 25·0°, 24·1°, and 21·9°. And hence the annual fluctuation of the mean monthly temperature was greater at 10 inches, about equal at 20 inches, and less at 30 inches, than that of the air.

The chief points in connection with the extreme monthly temperatures are—(1) while in several of the months the highest temperatures of the drained land were considerably in excess of the undrained land, in nearly half of the months the undrained land either equalled or slightly exceeded the drained land; but (2) the lowest temperatures of the undrained land were in every month lower than those of the drained land at 10 and 30 inches, and nearly in every month at 20 inches. Hence, the advantage of drainage consists more in keeping up the temperature of the land during periods of cold weather, than in favouring its increase during warm weather, when, indeed, it is not absolutely required.

We now proceed to consider the influence of drainage on hill pasture, and generally on the climate and rainfall of the country, concerning both of which there has been of late some anxious thought and discussion.

A growing temperature is the prime element of all vegetation. This must be secured by all means in our power, and one of the most effectual is proved to be drainage, by the interesting observations in question, which, as far as they go, may be considered a most important advance in true scientific farming.

The effect of drainage on hill pasture is much more decisive than in arable land. In arable land, at the depth of 10 inches, the mean temperature of the coldest month was, as stated in our previous article, 0·2° less than that of the air, and of the warmest month, 1·1° greater, whereas, in hill pasture, the mean temperature of the coldest month was 2·7° higher than that of the air, and in the warmest month the air and land were of equal temperature. Hence, the range of the monthly mean temperature was smaller in hill pasture, at the depth of 10 inches, than in the air, and still smaller than in arable land.

The details of the observations are as follows :—

The soil is composed of a mixture of moss and clay, with a clay subsoil. The land slopes to the east, but the portion of it on which the observations were made is level. On one side the drains were 1 foot deep, and uncovered, and on the other, 2 feet deep, and covered. The thermometer was fixed 6 feet from the nearest drain. On the 29th of November, the temperature of the air fell to 26·0°, being from 15° to 20° lower than it had fallen for some time, and the temperature of the drained land fell from 44·3° to 43·5°, and the undrained land from 44·0° to 42·5°, the drained land having thus fallen only 0·8°, and the undrained 1·5°. On each of the two next nights the temperature fell to 25°, and both underground thermometers fell 1·0° and then 0·5°, the drained land keeping constantly 1·0° above the undrained.

On the 9th of September, the air fell to 39°, being several degrees lower than it had been for some time, and the drained land fell from 52·3° to 52·0°, losing only three-tenths, whereas the undrained land fell from 52·0° to 51·2°, losing eight-tenths of temperature. On the next two nights, there was a further fall of the temperature of the air, and the fall in the drained land was 0·8° and 0·2°, and in the undrained land 1·0° and 0·2°. Thus, during these three days the drained land fell 1·3°, whereas the undrained fell 2·0°, the difference being seven-tenths. Besides, the cold first took effect on the undrained land, causing its temperature to fall more rapidly, but as it continued, the temperature of both soils came to fall equally.

So much for winter and autumn. We come now to the observations in spring. During the 24 days, beginning with the 2nd of February, the temperature of the air fell almost every night below frezing, the mean of the nights for the period being 30·1°. During this protracted frost, the drained land was 37·1°, and the undrained 36·7°, the drained land being thus nearly half a degree warmer.

On the 31st of July, the temperature of the air rose about 8° higher than it had been for some time; and on the 1st of August, the drained had risen from 50·5° to

51·5°, and the undrained land from 50·1° to 51·3°, the drained having thus risen 1·0°, and the undrained 1·2°. Similarly, on other occasions, when the temperature rose, the undrained land generally rose a little more than the drained land; the drained land, however, generally continued higher, though in several cases the temperature came to an equality in both soils.

The mean annual temperature of the drained land was 44·3°, and of the undrained land 43·9°, the drained being thus 0·4° warmer. Here, again, as in arable land, the year may be divided into two periods, the one from June to November, when the excess was large, being nearly 0·7°, and the other from December to May, when the excess was small, being scarcely 0·2°. These periods correspond in time with the similar periods in arable land, at the same depth, but the excess of the temperature of the drained pastures was not so large as in the arable land. The difference between the highest monthly extremes was generally less than the difference between the lowest monthly extremes, a peculiarity which was apparent also in the case of arable land.

Bearing in mind that the observations refer only to arable land of light sandy soil, under a rye-grass crop, and in the case of hill pasture, to clay soils mixed with decayed moss, the conclusions regarding the effect of drainage on the temperature of the soil appear to be as follows :—

1. The temperature of hill pastures and arable land is raised by drainage, the contrary result being doubtless caused by the cooling effect of the larger evaporation from the undrained wet land.

2. The temperature of hill pastures was not raised, by

drainage, to the same extent as that of the arable land. But, as suggested by Mr. David M. Home, the difference possibly arose from the nature of the soil in the two cases. The hill pasture on which the observations were made, was a clay soil, whilst the arable land consisted of light sandy soil. Such being the case, the question is: was the clay soil as thoroughly drained as the light sandy soil? If it was not, it is then not difficult to understand why the effect of drainage, in raising the temperature, should have been less manifested in the hill pastures. In clay soils, drains will not draw water as far as in light sandy soils; therefore, if in both cases the drains were at equal depths and distances, more water would be left in the clay soils than in the sandy soils, and more evaporation, with consequent loss of heat, would take place in the former.

3. During sudden falls of temperature, and during protracted cold periods, as when the soil is under a covering of snow, the cold finds its way sooner and more completely through undrained than through drained land. For undrained land, having the interstices between its particles charged with water to a greater extent than drained land, is less porous, and therefore a better conductor of heat to the outer air; and hence its temperature falls more quickly when the temperature of the air has fallen below it.

4. When the temperature of the air is higher than that of the soil, drained land receives more benefit from the higher temperature than undrained land. The reason probably is, that the drained land, especially if in a state of cultivation, is more easily permeated by the air.

5. When rain or sleet has fallen, the superfluous moisture soon flows away from drained land; hence, in

such circumstances, drained land possesses the great advantage of a comparatively equable temperature, whereas the temperature of undrained land is liable to considerable fluctuation, for, when soaked with rain-water, it is temporarily raised, and when soaked with melted snow, it is temporarily lowered.

6. Since, from the above causes, the temperature of drained land is in summer occasionally raised above undrained land, to the extent of 3·0°, often 2·0°, and still more frequently 1·5°, it follows, that the advantage derived from drainage is, in many cases, the same as if the land had been actually transported 100 or 150 miles southwards, and thus drainage must have the effect of altering the climate of a locality, and generally improving the climate of a country by raising the temperature of the land.

By raising the temperature of the land, the temperature of the air, as it flows over the land, will be raised by contact with the soil. This seems to be a sufficient explanation of this most important result.

The observations on the temperature of drained soil when under turnips, and of undrained soil when under old grass, brought out the somewhat unexpected result of the temperature of the drained land being less than that of the undrained. To understand this result, as observed by Mr. Home, it would be necessary to have explanations on various points. Were both soils in the same locality? Were both soils of the same nature? If both soils were in the same locality, and of the same nature, then the land under the crop of turnips was very differently circumstanced from the land covered by old pasture. Evaporation will take place much more rapidly from soil that is bare or partially so, than from soil which is·well clothed with grass. Heat radiates more rapidly from soil when covered only by turnips, especially if the leaves have decayed, than when covered by old grass. Snow melts more rapidly on a fallow or a turnip field, than on a grass field.

Mr. Home made some interesting experiments in illustration of these principles. He took two zinc boxes, 2 feet square and 1 foot deep, filling one of the boxes with sandy loam, and the other with strong clay. Each was suspended at the end of a balance, and the quantity of soil in each was so adjusted, that the one box was exactly equal in weight to the other. He then poured an equal quantity of water into each box, and placed a thermometer into the centre of each, immersed 6 inches into the soil.

Mr. Home found that the thermometer was almost always lowest in the sandy loam, and that before a week had elapsed, the sandy loam box rose above the level of the other box, in consequence, as he inferred, of more water evaporating from it than from the clay soil. Probably, in the former, the water was more uniformly diffused through the soil, so that there were more surfaces or points for evaporation than from the clay soil. In the clay box the water would be less interspersed or more accumulated.

The benefit of drainage to crops arises, perhaps, not so much from the higher temperature caused, although this is of vital importance to their safety in winter and spring, as from the improvement of the soil, by making it more *friable,* and by *drawing the air to the roots of plants.*

CHAPTER XIII.

THE INFLUENCE OF TREES ON CLIMATE, WATER SUPPLY, AND RAINFALL.

THE question of the influence of woods and forests in a country is equally interesting in the scientific and political point of view. Science is interested in them as important physical agents in creation; and political economy must take them into consideration if, as seems certain, trees in a country are in general absolutely necessary to secure its fertility,—to enable us to count upon average, if not superabundant, harvests for our increasing populations. In fact, it would seem that if we wish to increase the annual yield of wheat *we must plant more trees* throughout the country. Such is the positive upshot of the argument which we shall proceed to discuss.

In France a very considerable destruction of forests had taken place for a very long time, with the object of giving to tillage the land occupied by woods; but of late years the inundations which have devastated great extents of the country have compelled the Government to make a vigorous attempt to restore the forests on a great scale.

A vast deal has been said and written on the influence of trees in the general economy of nature, without elucidating the subject by reasoning founded on clear and precise experience; and therefore we must be thankful to any one who can furnish us with such downright facts as may enable us to draw our own conclusions.

The question comes home to us in England in connection with our decreasing rainfall and deficiency of water, complained of almost on all sides; and so, perhaps, it may be useful to give an account of the researches and conclusions of a French author, M. Becquerel, who has recently devoted his attention to it, and endeavoured to come to right conclusions on the subject.

The subject is twofold: in the first place, it relates to the social importance of forests; and, secondly, to the influence which they exert on the temperature and the humidity of the soil and the atmosphere, both of the latter being vital agricultural considerations.

The history of all nations reveals the causes and the effects of cutting down the woods and forests of countries. At first it was the increase of population which necessitated the measure progressively; but the great devastations only go back to the epoch when mighty conquerors, with the view of subjecting newly-formed nations, cut down and burned the forests which served as a refuge to the inhabitants. Therefore, the increase of populations, wars, and the progress of civilization have been the chief causes of the destruction of forests. The banks of the Euphrates, the Ganges, and the shores of the Mediterranean now present to the traveller only deserts and marshes, in which we scarcely find a trace of the exuberant tillage of past ages. From the time of Sesostris—far back in the ancient history of Egypt—

to that of Mahomed II., in the fifteenth century of the present era, Asia Minor was the theatre of devastating wars, which destroyed the forests; and the watercourses dried up. The Land of Canaan, which the Bible cites as the most fertile in the universe, is, at the present day, deprived of both vegetation and water, because her forests were destroyed. The beautiful forests, which formerly vivified the seaboard of Africa, are now at least one hundred and twenty miles distant from the shores of the ocean.

In the early history of France, we find that Julius Cæsar, in order to penetrate into Gaul, had to play sad havoc continually with the woods and forests. There, as all over Europe, wars and the progress of civilization have transformed vast extents of country into barren swamps, wastes, or regions of everlasting heath.

Contrary to the general assumption, it appears that those vast regions covered with forests did not render the climate so rude as is supposed to be the consequence of their existence, for Cæsar himself tells us that the territories in question were amongst the most fertile in the country.

The cutting down of the forests of France continued ever after; and in the ninth century the incursions of the Normans vastly contributed to the abandonment of tillage, and originated the wide wastes or "landes" of Brittany, and the deserts of Champagne and Poitou, which occupy the sites of the ancient forests.

From the time of Charlemagne many royal edicts were directed against this blind destruction of forests; but all such preservative measures left a great latitude for evasive interpretations, and the cutting down of forests has continued with deplorable facility. At length in 1859, more rigorous measures were taken. The law now permits the cutting down of forests to an extent of about 25 acres, only on the following conditions :— The land on mountains and slopes must be maintained ; the soil must be protected from inundations ; the springs must be preserved ; the downs must be preserved, and other precautions must be taken with reference to the public health. These restrictions would be sufficient if it were always easy to enforce them.

During the last sixteen years the extent of forests cut down in France has been at the rate of about 75,000 acres per annum. Against this there has been a re-planting of forests to the mean extent of about 25,000 acres per annum, the difference being 50,000 acres. The result is, that in the space of one century the French have lost *five millions* of acres,—that is, nearly one-fourth of the wooded surface of France, which is at present about 22½ millions of acres. Such is the state of matters in France, and that, too, in the face of the fact that France by no means needs such an increase of arable land, since the production of wheat there already exceeds the requirements of consumption.

Now, this clearing out of forests in France concerns us here in England more than we are apt to imagine, if, as seems certain, the existence of forests is directly connected with the *production of rain;* therefore, any interference with nature in this respect, as in all others, must sooner or later tell on her neighbours. We owe much moisture to our S. and S.E. winds, from their decomposition by the prevalent N.W. and W. and S.W. winds, in summer. Well, the vapour brought to us by S. and S.E. winds must have been generated in countries lying to the south and south-east of our island, as before demonstrated (p. 34). Any interference with nature is, in its effect, much like that of a pebble flung on the still surface of a lake, when the consequent circles go on for ever enlarging—Heaven only knows how far. Any disturbance in the rainfall of one country must necessarily react on that of its neighbour.

It may be a strange question to ask if it is in any degree within the power of man to alter the weather; but, as Sir J. Herschel observes, it is by no means so absurd as it appears. The total rainfall and—which is, perhaps, as regards weather and climate, of even more importance—the *frequency* of showers on an extensive well-wooded tract, or one entirely covered by forests, ought, on every theoretical view of the causes which determine rain, to be greater than on the same tract denuded of trees. The foliage of trees defends the soil beneath and around them from the sun's direct rays, and dispenses their heat in the air, to be carried away by winds, and thus prevents the ground from becoming heated in the summer; while, on the other hand, a heated surface-soil reacts by its radiation on the clouds as they pass over it, and thus prevents many a refreshing shower, which they would otherwise deposit, or disperses them altogether.

But this is not the only way in which man alters the weather. That admirable agricultural improvement, drainage, from having been unwisely carried out, has, so to speak, done as much harm as good to England's broad acres. By carrying away rapidly the surface-water down to the rivulet, and so hurrying it away to the ocean, drainage not only cuts off a great deal of the supply of local evaporation, which is a material element in the amount of rainfall, but, by causing the surface

to dry more rapidly under the sun's influence, it allows
it also to become more heated, and so to conspire with
the action of the denudation of trees to *prevent rain*.

Evidence is at hand to corroborate this theoretical
view of the matter. The rainfall over large regions of
North America is said to be gradually diminishing, and
the climate otherwise altering, in consequence of the
clearance of the forests; whilst, on the other hand, we
have the very remarkable fact that under the beneficent
influence of a largely-increased cultivation of the palm-
tree in Egypt, rain is annually becoming more frequent.
Again, there are lakes in what were formerly Spanish
America—that of Nicaragua, if we mistake not—whose
water supply (derived, of course, from atmospheric
sources) had become so diminished, owing to the denudation of the country under the Spanish *régime*, as to
contract their area, and leave large tracts of their shores
dry; but now, since the trees have become restored, are
once more covered by their waters. Even in our own
southern counties complaints are beginning to be heard
of a diminution of water-supply,—partly owing to gra-
dually-decreasing rainfall, from the universal clearance
of timber, although chiefly, perhaps, attributable to rob-
bing the springs of their supply by draining,—a practice
no doubt beneficial to agriculture if used with caution
and in moderation, but of which the consequences, if
carried to excess, may ere long be severely felt, in
rendering large tracts of country uninhabitable in summer from mere want of water. Our outfalls are every
day becoming more direct and rapid in their action.
Rain, which used to drag sluggishly downwards through
meandering streams, runs straight out to sea. Floods,
rather as the result of this arterial drainage than as the

effect of parallel subsoil drainage, follow excessive rainfalls more immediately than they used; and, unexpected
as it may be, the influence of agricultural drainage becomes apparent on our springs, our mill-streams, and
our rivers. It is of great importance that an inquiry
be made into this subject, in order that the truth of the
matter may be known before the owners of estates shall
have permanently injured their own interests by a too
narrow confinement of their views to the mere agricultural object of land-drainage. And still more is it necessary that the owners of land may not have permanently injured the general interests of the island and
its population before Government shall have lost any
opportunity of interference. Lastly, several discussions
have been held on the well-supply in and around London,
during which it has been elicited that a serious and pro-
gressive depression of water-level has been taking place
for a period of many years. Such have been the consequences of denudation and excessive or injudicious drainage,—the latter being a direct waste of precious water,
the former a prevention of the natural supply designed
by Providence.

Columbus, in his Journal, attributes to the extent of
forests which covered the mountains the abundance of
rain to which he was for a long time exposed whilst
coasting about Jamaica. He also remarks that the
abundant rains of Madeira diminished when the trees
were cut down. Humboldt shows that there exists
above wooded regions a frigorific radiation, which condenses the vapours. This may be rather incomprehensible; perhaps we should rather say there is an attraction towards the region of forests for the cirrus, or icy
cloud, which is indispensable for rain. The summits of

mountains covered with wood are more frequently enveloped with fogs than arid mountains.

In certain islands of the Antilles, the clearance of the
forests has diminished rainfall, and the watercourses
have failed. At Porto Rico, by a royal order of the
King of Spain, for every tree cut down three more must
be planted, and thus the great fertility of the island has
been maintained with abundant waters.

According to Boussingault, in the valley of Canca,
the cocoa-nut tree will not flourish near a forest; but as
soon as the forest is cut down the tree prospers; the
rains are too abundant, and prevent the ripening of the
fruit. This fact has been repeatedly observed.

White, of Selborne, says that in heavy fogs, especially
on elevated situations, trees are perfect alembics distil-
ling water; and no one who has not directed his attention
to such matters can imagine how much water one tree
will distil in a night's time, by condensing the vapours
which trickle down the twigs and boughs, so as to make
the ground below quite in a float. On a certain misty
day a particular oak in leaf dropped water so fast that
the cartway stood in puddles, though the ground in
general was dusty. In some of our smaller islands in
the West Indies, where there are no rivers or springs,
the people are supplied with water, merely by the dripping of large, tall trees, which, standing on the breast of
a mountain, keep their heads constantly enveloped with
fogs and clouds, from which they dispense their kindly
never-ceasing moisture.[1] Trees perspire profusely, condense largely, and check evaporation so much, that

---

[1] But this precipitation of fog by trees does not take place in a calm;
it requires a gentle wind.

woods are always moist, and, consequently, contribute much to ponds and streams.

The destruction of forests and trees in general exerts various influences on the climate of a country. This influence depends upon their extent, the height of the trees, and their nature,—their power of evaporation by their leaves, the faculty which they possess of getting warmed or becoming cooled by the air; lastly, on the nature and physical condition of the *soil and subsoil*. On the other hand, trees influence running waters and springs.

As a shelter from low winds the utility of woods is evident, it is proportional to the height of the trees. The evaporation which goes on by their leaves is a powerful and incessant cause of moisture; the least lowering of the temperature precipitates the vapours of the air, the resulting water penetrates into the soil, or is absorbed by the roots. On a soil denuded of trees it runs off, and adds to the bulk of watercourses, often causing inundations.

As to the calorific state of trees, or their capacity for receiving and parting with heat, it has been demonstrated that they get warm and become cool in the air, like all inorganized bodies, by the action of the sun. Being bad conductors of heat, they attain the equilibrium of temperature with the air only after a considerable time. The tree, warmed during an entire day by the sun's rays, gives off the greater part of the heat during the refrigeration of night, and restores by degrees to the surrounding air the heat which it has laid up. These facts have been verified by M. Becquerel by observations with the electric thermometer; and they refute the general assumption that trees tend to lower the temperature. On the contrary, it would appear that they increase it precisely at a time when it is most needed.

The influence of clearing out a country on the mean temperature has been studied by two eminent investigators, Boussingault and Humboldt. The former found, in the equinoctial regions of America, that the abundance of forests and humidity tend to make the climate cooler, whilst the dryness and aridity of the soil keep it warm. Humboldt, on the contrary, declared that in North America the climate has not been altered by the destruction of the forests; but was still of opinion that, in many parts, the clearance of forests would raise the mean temperature. The difference in the results of these two investigators is easily explained if we have due regard to the *nature of the soil*, which is of the utmost importance to the inquiry. M. Becquerel has specially examined the influence of denuded soil on the temperature. He has found that different soils exposed to the sun acquire temperatures varying from 43 to 54 degrees. In their refrigeration by radiation the difference is also very considerable. A siliceous or flinty soil cools more slowly than one in which chalk or clay predominates. Now, such being the case, it is evident that a clearance of forests from a flinty soil must raise the mean temperature more than a clearance from other soils, all other conditions being equal.

The effects of clearance on the springs and the quantity of running water in a country are a still more important consideration. Forests contribute to the formation of springs and river-sources, not only by means of the humidity which they produce and the condensation of vapour by refrigeration, but also by reason of the obstacles which they present to the evaporation of the

water in the soil itself, and by means of the roots of
their trees, which, by dividing the soil like so many perforations, render it more permeable and facilitate infil-
tration.   We have given numerous instances in proof of
these averments, and will only add one more.   The city
of Nueva Valencia, in South America, built in 1555,
was then situated about a mile and a half from a lake;
but by the year 1800 it was four miles and a half distant.   In 1822, however, the waters of the lake had a
rise, and the lands formerly cultivated were submerged.
The explanation is, that during twenty-five years the
valley in which the city is situated had been the theatre
of bloody conflicts, during the war of independence; the
population was decimated, the land had remained un-
cultivated, and the forests, which increase with such
prodigious rapidity in the tropics, had completely taken
possession of the country.   Such examples prove that
by clearing out a fertile country, depending for water
only on its springs, we run the risk of drying up the
latter to the extent of impoverishing the land.   By recent accounts in an American paper, it seems that even
the great Mississippi is "drying up" from the same
cause, and the time must come when it will no longer be,
as its Indian name signifies, "the mother of the waters."

In discussing the great question of the influence of
clearing out woods and forests on the water supply of a
country, M. Becquerel comes to the following conclusions:—1. Great clearances diminish the quantity of
running water in a country. 2. We cannot as yet decide
whether this diminution must be ascribed to a less annual quantity of water falling, or a great evaporation of
the rain-waters, to both these causes combined, or to a
new repartition of the rain-water. 3. The established

cultivation of an arid and exposed country dissipates a portion of the running waters. 4. In countries in which no change of cultivation has taken place, the quantity of spring-water seems always the same. 5. Forests, by preserving the spring-waters, economize and regulate their flow. 6. The humidity which prevails in woods, and the intervention of the roots in rendering the soil more permeable, must be taken into consideration. 7. The overflows of water in mountainous countries exert an influence on the watercourses and springs; in level countries they can only act on the springs.

Such is the various and complex influence of forests and trees in general on the climate of countries. If their clearance does not always entail sterility upon a country, it is injurious by facilitating the invasion of sand on the plains, as in France, where the winds drive the sands freely from the seashore, unopposed by the forests. With M. Becquerel, we may conclude that we ameliorate the climate of a country by clearing out heaths, draining and purifying marshy lands, and by planting with woods our mountains, hills, and other suitable localities; but that great clearances of woods and forests are neither necessary nor useful in any point of view whatever.

The fertility of soils is in proportion to their *absorbent power for water*; and therefore, if water be deficient, all "advanced" tillage, the most appropriate manures, etc., will be useless. The conclusion is, that we must reform many of our modern "advanced" notions about drainage, etc., and fall back upon the old methods of Mother Nature, who cannot go on for ever humouring us in our unwise experiments and interference with her rules and regulations.

CHAPTER XIV.

HOW TO PROGNOSTICATE THE SEASONS.

ALTHOUGH it is not in man's power to secure sunshine or completely to supply the deficiency of rainfall, still there are resources at our command to mitigate the consequences of these deficiencies, or to make amends for them by acting in accordance with the hints of nature. If a year is likely to be bad for grain crops, it will be good for roots, tares, and other food for stock. In such a case we only invite loss if we do not take means to secure a compensation by having recourse to this rotation, supposing we can discover the probability in time for the alteration.

Now there are numerous facts or classes of facts which enable us to foresee with more or less probability the nature of coming seasons.

I. THE WEATHER AT THE EQUINOXES.

The weather at the equinoxes (21st of September and 21st of March) is a sign of that which is to follow. If the weeks immediately before and after the autumnal equinox (21st of September) pass off almost free from any great atmospheric disturbance, *the temperature will continue higher than usual far into the winter months.*

Such being observed, it will be evident that the deficiency of frost will tell on the soil, and the irregular temperature will prepare unsettled weather for the following spring.   The contrary will happen under opposite circumstances.

On the other hand, a calm spring equinox will be followed by a very low temperature for many weeks; the reverse with a boisterous spring equinox, which means a dry spring.   This prognostic from the equinoxes was deduced by Dr. Kirwan, from a variety of meteorological observations made in England, between the years 1677 and 1788.   His inferences are as follows :—

(1.) That when there has been no storm before or after the spring equinox, the ensuing summer is generally dry, at least five times out of six.

(2.) That when a storm happens from any easterly point, either on the 19th, 20th, or 21st of March, the succeeding summer is generally dry, four times in five.

(3.) That when a storm rises on the 25th, 26th, or 27th of March, and not before, in any point, the succeeding summer is generally dry, four times in five.

(4.) That if there be a storm at S.W. or W.S.W. on the 19th, 20th, or 22nd of March, the succeeding summer is generally wet, five times in six.

This method of prognostication has been of late years revived by Mr. Du Boulay, in annual manifestoes on the coming harvest.

The wet and cold spring of the past year (1866) prognosticated a bad or inferior harvest,—as we ventured to predict in the ' Mark Lane Express ' during the summer.[1]   It would not be a waste of precaution, in certain

---

[1] The following was the opinion :—" The wet and cold spring of the present year (1866) prognosticates a bad or inferior harvest.   The rain-

seasons, to check the drainage, especially of high-lying lands, against the possibility of dry weather, if not drought, with thirsty winds to diminish the moisture of the soil.

For the great majority of years it seems the farmers cannot expect very remunerative harvests; but the quality of food for stock might be very profitably increased were we to alter the system of cropping, and grow less corn, the cultivation of which is both unprofitable and laborious, as observed by Mr. William Scott, at the quarterly meeting of the Galashiels Farmers' Club.

Though grain has got a little dearer this year, it is well known that it is entirely owing to circumstances which cannot be permanent, and it is only in years far distant, when the population of this world shall have doubled itself,—should that ever come to pass,—that there will be any chance of corn ever again rising to the old protection prices: wheat will never again be dear, and beef will never again be cheap. If more attention were paid to the cultivation of roots, tares, and other food for stock, the numbers kept might be greatly increased, and more especially would this be the case if more land were kept in grass. If these hints be taken, the farmer will not be so dependent on the weather as he is at present. By judicious drainage we can counteract the consequences of excessive rainfall; but then we should *save the water*—preserve it as a precious thing, against the possibility of having to pay for it at 6*d*. a pailful, as some had to pay during the last excessive

fall of the country seems already to have amounted to the greater part of the annual average more or less everywhere, whilst the general tendency of the winds is decidedly variable, denoting a continuance of unsettled weather."

drought. But even more than that: if, from the hints
we have to suggest, it be found that we are likely to
have a dry season, means may be taken to stop or check
the drainage, which should always be done in spring
after a wet autumn, especially in high-lying localities.

By a knowledge of the "climate" or natural temperature of localities, which depends upon their aspect with
respect to the sun, and the nature or geological qualities
of the subsoil, we should be able to neutralize the consequences of drought by the nature of the cultivation or
crop we raise from it.

Those plants which have broad leaves, such as roots,
etc., are not so likely to suffer from drought as the cereals, which is another reason why farmers had better
turn their attention to food for stock.

II. SIGNS OF THE HARVEST FROM THE WINDS.

Every wind has its weather, more or less influencing
vegetation, or interfering with agricultural operations.
Consequently this is a subject to which farmers should
direct their attention, and endeavour to turn to account
what little certainty we may have secured in our knowledge as to the play of the winds. Owing to the want
of continuous records of the changes of direction in the
wind, but little is known as to the mean prevalence of
certain directions during the different months and seasons. No doubt if self-registering anemometers (or instruments for measuring the force and direction of the
winds) were used throughout the country, this want
might be supplied, but even then the results would
scarcely amount to certainties, owing to the many causes
which deflect or turn out of their course the great currents of air as soon as they enter the varied landscape

of our islands. Besides, many a wind may blow under
a false name; that is, it may be only the return current
of one and the same wind,—sometimes in an exactly opposite direction. The only means of distinguishing such
false winds is, the hygrometer, whose capabilities we
have fully explained, and which will accurately show the
quality of any wind as to moisture, and therefore enable
us to class it accordingly with either of the great currents
to which it may belong, absolutely or in composition,
namely, the polar or the equatorial, which have their
marked and distinctive features. It is the relative prevalence of these two main currents that constitute the
winds which mainly influence vegetation.

Whatever, then, may be our difficulties in the inquiry,
it is highly desirable that the farmer should have regard
to the direction of the wind; for, if we cannot predict
for certain the class of winds which we are likely to experience during a certain period, we know that certain
winds have certain peculiarities of their own, as regards
their degree of humidity or moisture, dryness, tendency
to bring settled weather of some kind, whether good or
bad; and, therefore, these become very important elements for consideration in predicting the weather. A
good trustworthy wind-vane or anemometer,[1] which need
not be very expensive, would be a very useful thing to
the farmer, as will presently appear; at all events, he
should always have an eye to the direction of the wind
or its shiftings; for, after all, everything depends upon
the winds, as far as the crops are concerned. The deficiency or excess of one class of winds at particular
periods will forewarn us of results which might otherwise
meet us unawares.

[1] See Chapter XXIV., " Anemometry."

In a previous Chapter, p. 31, we have stated the average duration of each wind throughout the year at Greenwich. According to that summary, the south-west wind is by far the most prevalent, and the south-east the least so. The north and north-east rank next to the southwest winds in point of duration; and there are, on an average, only thirty-three or thirty-four days in the year on which the air may be said to be calm, although, strictly speaking, perhaps there are but very few days, if any, during which the air is perfectly motionless.

Of course, that general average can only serve to give some idea of the regular crop of winds that prevail in our islands. There must be differences in various parts of the country, owing to local causes; for instance, at Liverpool, the most prevalent winds are from S. to S.E., which last fifty-four days, and the N.E. and S.W. winds, which are our longest winds generally, only blow for twenty-four days at Liverpool, where they have the N.E. wind for only eight and a half days. We give the statement as we find it recorded by the observers at the Liverpool Observatory; but we submit that the very low and otherwise objectionable position of the observatory is scarcely adapted for accurate anemometry, and therefore it probably records only locally reflected winds.

But westerly winds mostly prevail in England,—that is, S.W., W., and N.W. winds; and it seems that this fact accounts for the tendency of the wealthier population to move westward or towards " the West-End," because the air there is freer from smoke, etc., the prevalent westerly winds not only driving back the smoke of the City, or " East-End," but carrying to it the smoke of the West-End. Of course, it is the same with exhalations, and so the westerly wind keeps the West-End healthier as well as makes it "respectable." We may observe that in east winds, owing to the greater density of the air, the smoke and exhalations of both East and West-End rise higher, and so each gets rid of a greater part of the nuisance.

Curiously enough, it is the same in other great cities, —Paris, Vienna, Berlin, Turin, St. Petersburg, Liége, Caen, Montpellier, and almost every other capital or large city of Europe, where the best districts are in the *west*, with the same results. Moreover, it was the same at Pompeii and other ancient cities, and the same may be observed even in small towns and villages throughout this and every other country in the northern hemisphere. Thus, if instinct did not lead to this preference, it may be satisfactory to know that for once, at least, fashion is right in its requirement.

According to the Astronomer Royal, as before stated, there are only eight points from which the wind ever blows steadily for any lengthened period in England. It never blows at all directly from due south; the two most prevalent winds are the S.S.W. and the W.S.W.; the S.S.W. invariably bringing rain, the W.S.W. accompanied by dry fine weather; between W. and N.W. there is another point of duration, also between N. and E., another between E. and S.S.W., which, with N., W., and E., make up the eight points as above stated.

This statement accords with our observation, and we believe it will be found generally correct, that the above-named winds S.S.W. and W.S.W., winds blowing between W. and N.W., between N. and E., between E. and S.S.W., and N., W., and E. winds are remarkable for their continuance.

Now, it is evident that the farmer can infer the consequences accruing to his crops with some degree of probability at the commencement of any of these winds. Those which blow from the left side of the compass will be more or less wet, whilst those that blow from the east side will be generally dry. In general, the stronger the wind, the longer it will continue.

The wind usually turns from N. towards S. by E. quietly, and without rain; but returns to the N. with a strong wind and rain. The strongest winds are when it turns from S. to N. by W.

After a long continuance of westerly winds, with wet, a shifting by the N. towards E. may be expected, between which points there will be a wind of some duration, with dry cold weather.

If the wind turns to N.E., and continues two days without rain, and does not turn S. the third day, and there be no rain on the third day, we may infer its duration, and it is likely to continue N.E. for eight or nine days, all fair, and then to come to the S. again.

Fair weather for a week, with a southern wind, is likely to be followed by a great drought, if there has been much rain out of the S. before.

If, in unsettled weather, the wind veers from S.W. to W., or N.W. at sunset, expect finer for a day or two.

It is a very old saying, that when the wind turns against the sun, the contrary way to the hands of a watch, the weather will be wet and stormy. Whether stormy or not, it will then be unsettled, and thus be prejudicial to the crops.

Now, this well-established maxim, besides being applicable to any period of time, seems to apply equally to the entire year. It will, we believe, be found, without hardly an exception, that when the number of these re-

trograde or backward revolutions have nearly equalled
or exceeded the revolutions of direct motion in any year,
that year has been characterized by stormy and bad seasons; and that the opposite effect is observed to take
place when there has been a large preponderance of direct
over retrograde motion.   From 1849 to 1861 inclusive
(thirteen years), the vane made one hundred and sixty-
six complete revolutions more in the direct than in the
retrograde motion, or, on an average, nearly thirteen revolutions per annum, which shows the sequence or regular shifting before given.   In all this interval, two years
only, 1853 and 1860, gave a contrary result, and that
only to the total amount of two revolutions in excess the
wrong way in each.   Now, the harvest of 1853 was certainly below the average, and so was that of 1860, which
was especially characterized by stormy weather; so was
1846; and we may expect the same next year (1867), a
bad or inferior harvest, unless, as before observed, our
extensive drainage comes to the rescue.

In other words, the *minimum* of direct motion appears
to be reached at every seventh year, which may be termed
the Sabbath of the land, when " it rests from its labours."

The maximum of direct motion, indicative of good
years for the harvest, occurs at periods equidistant from
those of the minimum, or at three and a half years from
each minimum, and still seven years apart.   Thus 1863
was a good year, the crops being returned everywhere
as " over the average," " above the usual yield;" in fact,
a maximum harvest, the like of which we may expect in
1870, the intermediate harvests being either average or
below it; and this, we think, is indicated by the beha-
viour of the vane—that is, the rotation of the wind—

from January to June.   Continued backing of the wind during spring is a bad prospect for the harvest; for it means unsettled weather.

The following is the record of Osler's anemometer from the year 1841 to 1860 (inclusive), in which year the wind made rather more than two retrograde rotations, or in the direction of N., W., S., E., N.   The letter D denotes when the wind turned in the direction N., E., S., W., N.; and R when in the opposite, or N., W., S., E., N. :—

The Vane made, in the Year	*Revolutions.*
1841	D 5·4
1842	D 13·1
1843	D 20·7
1844	D 21·7
1845 (10½ months)	D 8·9
1846	D 1·8 (minimum)
1847	D 11·0
1848	D 12·1
1849	D 23·3
1850	D 15·9
1851	D 19·1
1852	D 8·8
1853	R 1·9
1854	D 6·8
1855	D 10·8
1856	D 16·1
1857	D 14·7
1858	D 24·1
1859	D 14·0
1860	R 2·1

The Astronomer Royal suggests that the explanation of this septennial cycle, or period, must be in a "periodical throb of temperature from the interior of the earth;" but, supposing that wind-currents are influenced generally by temperature and by the action of agencies *exterior* to the earth, the fact of "a periodical throb" from

the interior of the earth would hardly seem to be a true
cause for changing air-currents (unless it were local, like
a volcanic eruption), if stated by any ordinary authority;
but, coming from Mr. Airy, it must, of course, be taken
to be important.

As to particular times, it has been observed that in
the space of fourteen days, or half the lunation, the winds
ordinarily make an entire revolution, and blow successively from all points of the compass. At new moon, if
the wind be north or northerly, it passes on to the east
in a few days, then to the south, and so on to the west,
and returns to the north about the full moon, with settled
weather. In unsettled seasons, as before remarked, the
winds will often vary, and run a little backward, apparently against the course of the sun; but they seldom
veer quite round in this manner, stopping at some intermediate points.

The following table gives the relative prevalence of the
different winds during the year, in days and decimal
parts of a day, as observed at the Royal Observatory at
Greenwich; and it may serve to give an idea of their
play, if nothing more:—

	N.	N.E.	E.	S.E.	S.	S.W.	W.	N.W.	Calm.
January . .	3·0	3·3	0·8	2·2	4·1	9·8	3·5	1·5	2·8
February . .	3·0	3·6	2·1	1·3	3·0	8·0	2·9	2·0	2·6
March . . .	3·7	4·1	2·5	2·1	2·6	7·7	3·3	2·8	2·4
April . . .	4·0	6·1	3·5	2·1	2·4	6·3	2·6	2·3	1·0
May . . .	4·3	7·0	2·5	1·7	2·7	7·6	2·0	1·3	2·0
June . . .	3·3	3·6	2·2	1·7	2·2	9·9	3·7	2·1	1·6
July . . .	3·4	3·7	1·2	0·6	2·7	10·6	4·0	2·5	2·6
August . . .	3·0	3·0	1·2	1·3	3·0	10·5	3·9	2·0	3·4
September .	3·6	5·3	1·9	1·7	2·0	7·3	2·6	1·5	4·4
October . .	3·1	2·5	1·2	1·8	3·4	9·1	4·3	2·1	3·7.
November . .	3·8	3·6	2·1	2·0	3·4	7·7	2·1	2·2	3·3
December . .	2·7	2·1	1·8	1·8	3·0	9·9	3·9	2·1	4·1

As to the most important matter, whether we are likely to suffer from drought or not in any year, we believe that there is no chance of drought if the winds on the left-hand side of the compass, including south, do not greatly predominate beyond the number of days recorded in the above table from December to March.

Steady winds from the right-hand side of the compass during March, in excess of those from the left, will never, we believe, be followed by drought, but, on the contrary, by genial weather and an abundant harvest. " A dry, cold March never begs bread." " A wet March makes a sad autumn."

If the wind be east or north-east in the fore part of the summer, the weather is likely to continue dry. If the wind be westward towards the end of the summer, then will it also continue dry. In low-lying lands, but with a good sunny aspect, a cold and windy May is a good prospect for the harvest; hence the special proverb :—

> " A cold May and windy
> Makes a full barn and findy."

Much wet in May is worse than excessive drought: hence the proverb :—

> " A May flood
> Never did good."

If the last eight days of February and the first twenty days of March are for the most part rainy, then the spring and summer quarters will be so too. It is said that a great drought always enters at that season. Obviously, this axiom is connected with the observation respecting the spring equinox before-mentioned; but we believe that the boisterous or calm weather of that period will be found to be a safer sign than wet only.

Generally, a moist and cold summer portends a hard winter.

A dry, hot summer and autumn, especially if the heat and drought extend far into September, portend an open beginning of winter, and cold succeeds towards the latter part and beginning of spring.

A warm and open winter portends a hot and dry summer, or if not hot, at all events a dry one.

Birds that change countries at certain seasons, if they come early, show the temper of the weather, according to the country whence they come; as, in winter, woodcocks, snipes, field-rakes, etc., if they come early, show a cold winter; and the cuckoos, if they come early, show a hot summer to follow.

A serene autumn denotes a windy winter; a windy winter, a rainy spring; a rainy spring, a serene summer; a serene summer, a windy autumn; so that the air, says Lord Bacon, " on a balance, is seldom debtor to itself; nor do the seasons succeed each other in the same tenor for two years together."

In the absence of an anemometer, farmers should note the shifting of the winds, as shown by an ordinary vane or weathercock. Care should be taken, however, that the vane be placed in as open and elevated a position as practicable, so as not to be influenced by indirect or reflected currents of wind. Nothing is steadier than an elevated vane, such as that at the top of the Cattle Market at Islington, and we wonder why that elevation, or the top of one of the great hotels so long empty, has not been converted into a regular meteorological or weather-observatory for the guidance of the farmers. Furnished with the usual few instruments required, and attended to by a competent observer, such an institution would be of great aid to those who are striving to collect the facts from which they have to answer the numerous inquiries made to them by farmers and others from all parts of the kingdom. If the proprietors of the Cattle Market will only take our hint and put it into execution, there can be no doubt that in a few years the science in which they and all of us are so much interested, as connected with the greatest of our material interests, will be vastly promoted. Besides what is going on below near the ground, it would be useful to ascertain what is taking place in the current of air some seventy or eighty feet above, or at a point not so likely to be influenced by casual causes of error. Changes of weather begin from above downwards.

CHAPTER XV.

THE CHARACTERISTICS AND THE METEOROLOGY OF THE SEASONS.

I. WINTER.

> " In sable cincture, shadows vast,
> Deep-tinged and damp, and congregated clouds,
> And all the vapoury turbulence of heaven,
> Involve the face of things.   Thus WINTER falls,
> A heavy gloom oppressive o'er the world."

ENDING and beginning the year, Winter is the sleep of nature, in long, dark, and dreary nights, which, in the depth of it, seem continuous,—the short and gloomy days, for the most part, only serving to make " darkness visible."   Snow, rain, and all sorts of disagreeable weather are its attendants; its only relief being the genial frosts which exert so beneficial an influence on the soil, preparing it for the production of an abundant harvest.

The leaves, for the most part, are fallen; but still there is no absolute lack of adornment in the garden and the fields. The Sweet Coltsfoot (*Tussilago fragrans*) blooms in our gardens, and scents the air for a long way all around with a smell that reminds us of some of the sweetest spring flowers.   In general, however, there is nothing interesting to be seen in the vegetable kingdom but lichens, mosses, and the bright berries of some of our shrubs and evergreens.

The principal berries which adorn the country on the naked boughs during the winter months are as follows :—

The Holly (*Ilex Aquifolium*), with scarlet berries.

The Ivy (*Hedera Helix*), with berries green.

The Pyracantha (*Mespilus Pyracantha*), with berries bright orange.

The Whitethorn(*Cratægus Oxyacantha*)with berries red.

The Blackthorn (*Prunus spinosa*), with bluish-grey berries.

The Bittersweet Nightshade (*Solanum Dulcamara*), with red berries.

The Misletoe (*Viscum album*), with berries white.

The Yew (*Taxus baccata*), with berries red.

These and several other shrubs bearing ornamental berries should find a place in our gardens; they adorn nature when all but the evergreens are leafless, and serve to decorate our windows and churches at Christmas.

All lovers of trees and birds should plant the holly freely; it is valuable for its shelter during wintry gales, for its glossy leaves, and its bright scarlet berries,—for its moral and poetical associations.

But it must be admitted that the festivities of Christmas are no longer what they were. The day of Christmas gambols, Christmas-boxes, etc., seems to have set with the rise of the era of railroads, and all the good old customs are gone for ever. There was one, however, the abolition of which can scarcely be regretted: an old custom prevailed of bleeding and sweating horses on the 26th of December, "for good luck to them through the year." It is the church-festival of St. Stephen, the first martyr, and perhaps the ingenious will find some connection between that fact and the torture of the horses.

Those two beautiful and well-known birds, the field-fare and the redwing, are the first winter visitants which attract our attention. At first we meet with little parties in the meadows and pastures, and by hedgerows, where the black berries of the elder and the ruddy haw hang in thick profusion.

The birds are rather weakly and comparatively tame; but by-and-by they become stronger, and assembling in large flocks, they chiefly haunt open fields, until the nights become frosty, when they breakfast on the berries of the yew, holly, hawthorn, and ivy—in their season,—and withdraw to the fields when the ground becomes thawed. As a general rule they seldom feed entirely on these berries, except during hard frosts and snowstorms; it is then that the low wailing chirp of the redwing is most heard, and seems expressive of deep distress.

The berry-loving propensities of the missel-thrush are much stronger than either of the two first-mentioned birds, and his very quarrelsome disposition never fails to manifest itself against all birds, both great and small, which happen to feed in his company.

The gentle song-thrush migrates from many inland districts to the seacoast on the approach of winter, and during the hardest weather he gleans industriously his favourite food of snails, singing particularly loud at the approach of rain;—hence called the *Storm Cock*.

Another attendant of winter is the wary blackbird, which rarely ventures far from the shelter of bush or hedgerow, delights in fruits, cultivated, if he can get at

them, or wild, whenever they can be procured. During
very hard weather he may be seen in the rick-yard
eating grain, or filching from the pigs' trough ; but
where full-grown hollies exist, it is a pretty sight to see
this mellow songster picking the bright berries with his
coral bill, amidst the falling snow.

We have one well-known sign of the approach of
winter in the increasing familiarity of Master Cock
Robin—poor Cock Robin, "sitting on a rail "—the
loved of all for the place which he fills in the legendary
lore of the nursery, which has given him a place in the
affections of both old and young, to which his excessively quarrelsome disposition gives him no good title.
But, after all, this evil extends not beyond the society
of his fellows ; in the company of man he often evinces
the most engaging familiarity ; he attends the gardener's spade, enters our churches and our dwellings,—a
favourite everywhere ; for there is no withstanding the
wistful glance of his full black eye.

With regard to its meteorology,[1] our winter begins,
by the temperature, on the 7th of December, and continues eighty-nine days,—in leap-year, ninety days.

The mean temperature of the twenty-four hours, in
London, descends from about $40°$ to $34\frac{1}{2}°$, and returns
again to the former point.

The mean height of the barometer is $29·80$ in., being
$0·02$ in. above that of autumn. The range of the mercurial column is greatest during this season ; and in the
course of twenty winters it visits nearly the two extremities of the scale of three inches. The mean winter
range is, however, $2·25$ inches.

The predominant winds at the beginning of winter

[1] Daniell's estimates differ a little from those of Luke Howard.

are the S.–W. In the middle, these give place to
northerly winds,—after which they prevail again to the
close.   They are at this season often boisterous at night.

The mean evaporation of winter, taken at situations
which give more than the natural quantity from the
surface of the earth (being 30·46 in. for the year) is
3·58 inches. This is a third less than the proportion
that should correspond with the mean temperature,—
showing the *dampness* of the air at this season.

The average humidity by the hygrometer is 91, saturation being taken at 100.

The average rain of winter in London is 5·86. in.
The rain is greatest at the commencement, and it diminishes in rapid proportion to the end.   In this there
appears a salutary provision of Providence; for, were it
to increase, or even continue as heavy as in the autumnal months, the water, instead of answering the purpose
of irrigation, would descend from the saturated surface
of the higher grounds in perpetual floods, and waste
the fruitful plains and valleys. This sometimes happens, however, as during the winter of 1808, after a
frost and deep snow.   " On the 24th of January a steady
rain from west decided for a thaw. This and the following night proved stormy ; the melted snow and rain,
making about two inches depth of water on the level,
descended suddenly by the rivers, and the country was
inundated to a greater extent than in the year 1795.
The river Lea continued rising the whole of the 26th,
remained stationary during the 27th, and returned into
its bed in the course of the two following days. The
various channels by which it intersects this part of the
country, were united in one current, *above a mile in
width,* which flowed with great impetuosity and did

much damage. From breaches in the banks and mounds, the different *levels*, as they are termed, of embanked pasture land, were filled to the depth of eight or nine feet. The cattle, by great exertions, were preserved, being mostly in the stall; and the inhabitants, driven to their upper rooms, were relieved by boats plying under the windows. The Thames was so full during this time that no tide was perceptible; happily, however, its bank suffered no injury; the evacuation of the water from the levels proceeded, in consequence, with little interruption, and was pretty fully effected by the 23rd of February, or one month after. No lives were lost in these parts. Several circumstances concurred to render this inundation less mischievous than it might have been, from the great depth of snow on the country; it was the time of *neap* tide,—the wind blew strongly from the *westward*, urging the water *down* the Thames; to which, add moonlight nights and a temperate atmosphere, both very favourable to the poor, whose habitations were filled with water."

The public papers of the time give numerous details of inundations consequent on the thaw in question, which appear to have prevailed in low and level districts all along the east side of the island, but in no part with more serious destruction of property, public works, and the hopes of the husbandman, than in the Fens of Cambridgeshire, where, by some accounts, sixty thousand, by others above one hundred and fifty thousand acres of land were laid under deep water, through an extent of fifteen miles.

The following fact is worth preserving :—" About five hundred sacks were filled with earth and laid on the banks of the Old Bedford River, at various places where the waters were then flowing over. This proved effectual in saving that part of the country from a general deluge."

Nothing is more common during our winter in London, than to experience considerable indications of moisture in the intervals of our short frosts; and yet the actual quantity of aqueous vapour in the atmosphere is then probably at its lowest proportion ; or rather, it is so at the commencement of the season, after which it gradually increases with the *temperature and evaporation.*

This reduced state of the vapour influences the appearance of the skies. Clouds are formed of vapour, and therefore the less vapour the less the display of cloud-land. The shapes or forms of clouds depend upon their electricity, and as this element is generally weak in winter, the clouds exhibit little variety. On account of the low temperature they are easily, and therefore frequently, resolved into rain. The chief kinds that appear are those termed *cirro-stratus,* forming what is more significantly called by the people, " a mackerel sky," and the *cumulo-stratus,* the sort which must have been pointed to by Hamlet in the play, as being " very much like a whale," clouds of gigantic fancy forms, such as towers, huge rocks, scenes of battle, fantastic monsters.

Concerning the first, the popular adage is :—

> " Mackerel's scales and mares' tails
> Make lofty ships carry low sails ;"

and with respect to the second :—

> " When clouds appear like rocks and towers,
> The earth's refreshed by frequent showers."

Our winter sky hangs low, and the region below it to the earth is more or less misty. Yet we are not wholly exempt from thunder-storms during winter; they occur apparently in consequence of the sudden and plentiful decomposition brought on by strong southerly winds.

Hail is, however, of rare occurrence in our winter, if we except a sprinkling of small opaque *grains* which in the fore part of the night *indicate the approach of a low temperature*, and are found on the frozen ground, and on the ponds in the morning.

The snow crystallizes with us, when slowly and scantily produced, in forms not so various perhaps as those of higher latitudes, yet sufficiently beautiful to be worthy at all times of examination. The "star" of six rays, carrying more or less of secondary branches at an angle of 60°, is the most common.

After a copious fall of snow, an observer may find in the scenery which it forms, matter on which to exercise his powers of reflection. The pensile drifts, which in a mountainous country are objects of just alarm, may be contemplated here in safety, to discover the principles of their construction, and the manner in which they rest on so narrow a base. When the sun shines clear and the temperature is at the same time too low for it to produce any moisture, the level surface may be found sprinkled with small polished *plates of ice*, which refract the light in colours as varied and as brilliant as those of the drops of dew. At such times, there are also to be found on the borders of frozen pools, and on small bodies which happen to be fixed in the ice and project from the surface, groups of feathery crystals of considerable size, and of an extremely curious and delicate structure.

From the moment almost that snow alights on the ground, it begins to undergo certain changes, which commonly end in a more solid crystallization than that which it had originally. A notable proportion evaporates again, and this at temperatures far below the freezing-point. This evaporation from snow may very well supply the water for forming those thin mists which appear in intense frost; and it appears that the air in a still frosty night becomes partially loaded, either with spiculæ of ice or with particles of water, at a temperature below freezing, and ready to become solid the moment they find support. Hence the rime on trees, which is found to accumulate chiefly on the windward side of the twigs and branches.

A curious fact respecting snow may here be stated. The flakes of soot which are deposited on the surface of snow, and remain there exposed to the sun's rays, disappear after some hours, leaving a cavity, the bottom of which is visible and clean. There is therefore, probably, a real oxidation of the carbon, after which it is dissolved in the water, in the way in which the colouring matter of cloth is destroyed in bleaching.

The *Rime*, also, which collects on our trees and shrubs, when it just freezes with a moist air, presents considerable variety, and is occasionally magnificent. The *Hoar-frost*, which whitens our fields, usually at the approach of rain, of which it is a prognostic,—and is not confined to this season,—is of two kinds. The most common is *spicular*, like rime, and collected in this form from the air; though it may be doubted whether the particles are usually frozen until the moment of their attachment to the support; the other is *granular*, and consists of the drops of dew, beautifully solidified by the cold as they rest on the herbage.

Our great frosts are preceded by continued thick *mists*, arising from the condensation of the vapour which continues for some time to be emitted by the river and other waters, as well as by the moist soil, until frozen to some depth; for it is contrary to experience to suppose that the frozen state of the surface can prevent the ascent of vapour from the porous soil below, which will continue to emit it until its temperature becomes, by the gradual penetration of the frost, nearly on a level with that of the cold air then constantly flowing over it.

The simple difference of 4° or 5° in the mean temperature, suffices sometimes to effect the change from a damp misty state of the air to comparative dryness and serenity,—or the contrary. Our winters, therefore, present every variety of weather which can be expected within the limits of the temperature; from the calm frosty night, with its short day of cheerful sunshine, to the gloomy or thickly clouded sky, when the south-west wind surges among the leafless trees through the nights, or the more dreaded north-easter prevails through the twenty-four hours, driving the snow before it.

From the uncertain occurrence of really dangerous weather in our winters, it is probable that the people make less of the needful provision of clothing, use less foresight in their movements, and, in effect, suffer more in proportion from the cold, than the inhabitants of higher latitudes. That this improvidence exists is proved by the fact that drapers, outfitters, and other like tradesmen, run a good business at the first appearance of severe frosty weather, when everybody is reminded of his deficiency of warm underclothing. Amongst our severest winters, the memory of which may be fast fading away by reason of our late mild seasons, may be mentioned that of 1813–14, when "the Thames was frozen over," and whose approach was similar to that of 1867.

The frost began on the 27th of December, 1813, and by the 12th of January, 1814, the river Lea was firmly frozen and the Thames became so much encumbered with ice that navigation was scarcely practicable; by the 14th, the masses of ice and snow had accumulated in such quantities at London Bridge, on the upper side, that it was impossible for barges or boats to pass up, and by the 20th, that of the Thames below Windsor Bridge, called Mill River, was frozen over, and crowded with persons skating. It was not, however, before the 2nd of February, that the confidence of the public in the safety of the passage over the frozen surface of the river was established. On that day, all the avenues from Cheapside to the different stairs on the banks of the river were distinguished by large chalked boards announcing "a safe footway over the river to Bankside;" and, in consequence, thousands of individuals were induced to go and witness so novel a spectacle, and many hundreds had the foolhardiness to venture on the fragile plain, and walk, not merely over, but from London Bridge to Blackfriars. Several booths, formed of blankets and sail-cloths, and ornamented with streamers and various signs, were also erected over the very centre of the river, where the visitors could be accommodated with various luxuries. In one of the booths the entertaining spectacle of a whole *sheep roasting* was exhibited. Several printers brought presses and pulled off various impressions, which they sold for a trifle; here is a copy of one:—

"Printed to commemorate a remarkably severe frost, which commenced, December 27th, 1813, accompanied by an unusual thick fog, that continued eight days, and was succeeded by a tremendous fall of snow, which prevented all communication with the northern and western roads for several days. The Thames presented a complete field of ice between London and Blackfriars Bridges, on Monday, the 31st of January, 1814. A fair is this day (February 4th, 1814) held, and the whole space between the two bridges covered with spectators."

This "field of ice" was indeed a very rugged one, consisting of masses of drifted ice of all shapes and sizes, covered with snow, and cemented together by the freezing of the intermediate surface. The deceitfulness of the latter caused (as is too common on such occasions) the loss of some lives by drowning. The following passage, announcing the opening of the river soon after, is worthy of preservation :—

"*February* 11*th.*—We are happy to see the lately perturbed bosom of Father Thames resume its former serenity. The busy oar is now plied with its wonted alacrity, and the sons of commerce are pursuing their avocations with redoubled energy. Cheerfulness is seated on the brow of the industrious labourer; those who were reduced to receive alms as paupers, again taste the sweets of that comparative independence with which labour crowns the efforts of the industrious. What a fruitful source of congratulation does this prospect afford! Nor can the contemplative mind dwell on the subject without feeling gratitude to that beneficent Being, who, in time of such calamity, opened the hearts of the benevolent to administer, from their abundance, to the necessities of their poorer brethren, and thus add cement to the bond by which all mankind are linked together."

The mischief done on the river during this frost was greater than could be remembered by the oldest man living. Among the craft alone, it was calculated to amount to upwards of £10,000, independent of the damage sustained by the cables, tackle, etc., of the shipping.

We can spare space for only one more instance of severe frost, and that, on account of its connection with one of our weather-prophets, the famous Murphy, of almanac celebrity.

It was in 1838, the very first year in which Murphy published his 'Weather Almanac,' and his announcement for the 20th day of January was:—

" Fair.—Probably lowest degree of winter temperature."

By a lucky chance for him, this proved to be a remarkably cold day.   At sunrise, the thermometer stood at 4° below zero ; at 9 A.M. it was + 6° ; at 12 (noon), + 14° ; at 2 P.M., 16½°, and then increased to 17°, the highest in the day,—the wind veering from east to south.

The lowest degree of temperature for the season seemed to be reached ; and the supposition was proved by other singular circumstances, particularly the effect seen in the vegetable kingdom.   In all the nursery-grounds about London the half-hardy shrubby plants were more or less injured.   Herbaceous plants alone seemed little affected, in consequence, perhaps, of the protection they received from the snowy covering of the ground.

Two things were most remarkable, as being almost unprecedented in the annals of meteorological observation in this country ; first, the thermometer below zero for some hours, and secondly, a rapid change or range of nearly fifty-six degrees.

What a hit was this prediction for a weather-prophet ! The publishers were beset, the printers could not " work off" copies enough to supply the ever-increasing demand,

and it is quite certain that with our modern printing resources that *furore* after ' Murphy's Almanac' might have been made to realize some £15,000 or £20,000. As it was, Murphy cleared £7000 as his share for that year's almanac. He lost it all, however, not long after in some unfortunate speculation, showing that weather-wisdom and worldly wisdom are two very different things.

We may observe, however, that there was nothing very remarkable in Murphy's indication, because the coldest day in the year is generally about the time he specified ; January 20, or, at any rate, the first fortnight of January is almost always the coldest in the year, and so the probability of Murphy's coldest day coming true was owing to his placing it immediately after that usual term of greatest cold, in the event of frost continuing. With the view of showing that it was the merest chance that the grand prediction in question came true, the ' Book of Days' gives the following results :—

	Partly right on days	*Decidedly wrong* on days
January	23	8
February	8	20
March	11	20
April	15	15
May	12	19
June	18	12
July	10	20
August	15	15
September	15	15
October	11	20
November	14	16
December	15	16

So much for almanac predictions; but we may remark, that, by the above showing, Murphy was decidedly nearer the mark than most of his modern successors.

P

It should be stated, however, in fairness, that the *thaw* which followed this severe frost was also predicted by Murphy, if we are correctly informed, as to occur at or about the 6th of February. At any rate, our informant well remembers Murphy's almanac prediction of the thaw in question, and its occurrence about the time specified. In the ' British Almanac and Companion' we find the following statement :—

" Feb. 8th, 1838.—The navigation of the river Thames, which had been suspended in consequence of the frost having blocked up the river with ice for some weeks, was this day renewed, the thaw of the last two or three days having sufficiently cleared the river. So severe a winter had not been known for many years."[1]

II. SPRING.

" Be gracious, Heaven! for now laborious man
Has done his part. Ye fostering breezes, blow !
Ye softening dews, ye tender showers, descend !
And temper all, thou world-reviving Sun,
Into the perfect year !"

As early as Candlemas Day, the 2nd of February, the gloom of winter is mitigated ; the increasing day is now sensibly longer, and the fact is discovered by there being no need of gaslight or candlelight until near six o'clock. Some remark it by being able to take their tea by daylight. The weather is generally milder, and the exception to this rule—or a frosty Candlemas Day —is found so generally indicative of a cold early spring, that it has given rise to several proverbs, among the rest :—

---

[1] Steinmetz, ' Manual of Weathercasts.'

"If Candlemas Day be fair and bright,
Winter will have another flight."

The flowering of the Snowdrop is the harbinger of spring, although, as Mrs. Barbauld fancifully says, "Winter still lingers on its icy veins." It rises above the ground and generally begins to flower by Candlemas Eve,—"the first pale blossom of the early year." The Yellow Hellebore, or Flower of St. Paul, accompanies, and even sometimes anticipates the Snowdrop, and lasts longer, mixing agreeably its bright sulphur-yellow with the deeper and orange-yellow of the Spring Crocus, which blows, on an average, about the feast of St. Agatha, February 5th, and continues throughout March, fading away before Lady Day.

Snowdrops, Crocuses, and Hellebores give a well-kept garden a very brilliant appearance in early spring, whilst, as it advances, the continual increase of sunshine, not without the genial showers, brings to view other floral beauties of nature. The Yellow Coltsfoot blows, early Daffodils, and the large early Jonquils adorn our gardens, and in some places the former covers whole fields with its pale yellow; Daisies are seen in the fields; the sloping glades and shady banks and meadows are soon spangled with the little golden stars of the Pilewort; the Sweet Violet blooms in our gardens, and its rich odour perfumes our path before the clump of deep blue flowers from which it issued is discerned—like Charity, which lets not its left hand know what its right does for poor humanity.

But it is the animated creation that the sunshine of spring wakens into renewed life and activity, such as to give the season its distinctive character. It is the pairing time of birds, and then the earlier songsters warble their nuptial songs,—the Chaffinch, the Redbreast, and the Wren. The Throstle, the Missel-thrush, and the Blackbird are singing, and the Woodlark renews his note. Bulfinches return to our gardens, and are very useful, destroying those buds alone which contain the larvæ of destructive insects. The loud, shrill laugh of the Yaffle, or Green Woodpecker, "tapping the hollow beech-tree," is heard in the woods. Pigeons coo, and by this time usually have young. The Raven lays, and the Crow soon follows her in the labour of incubation, while the busy Rooks, already returned to their old nests, are heard, especially in the morning, while employed in repairing the damages of winter to their frail tenements. Frogs are now heard croaking from the ponds, ditches, and other waters; Snails are found clustered on the warm south walls, by the early blossoms of the Peach-tree; Toads make now a grunting noise, and the Stone Curlew, which arrives during the last days of February, is now heard by night, flying overhead unseen, and uttering its harsh, shrill cry. Owls then begin to hoot, and continue throughout the season.[1]

[1] Every bird has a note or a modulation of notes peculiar to himself, yet, what seems extraordinary, many birds decidedly imitate the notes of others. The Blackcap and the Thrush mock the Nightingale; and hence it happens that in the north and west of England, where Nightingales do not abound, the notes of these mocking songsters is less musical and less varied. Many other birds mock the Nightingale, and also mock each other. According to Forster, the average days on which birds arrive may be found out by the naturalist from their notes as well as by seeing them; and to those skilled in the music of the grove, this forms a very pleasant amusement during the bright fine weather of a spring morning. Forster knew persons who could distinguish the notes of every bird in the garden immediately on hearing him, but who, at the same time, were so little favoured by Apollo with regard to common music, that they could not tell "Rule Britannia" from the "College Hornpipe."

On fine days, towards Lady Day, the early Sulphur Butterfly—the emblem of the soul—is seen fluttering in the sunbeams; and the Bee—emblem of the mortal worker—goes abroad to its labours.

At a more advanced period of this season, and soon after Lady Day, the red and yellow Crown Imperial and the Dog's-tooth Violet blow; the Primrose and Dog Violets, which have blossomed sparingly before, now cover every bank and brae in profusion, and mix agreeably together. These plants extend their flowering into the beginning of the next season, and are scarcely out of bloom by the 24th of May,—a day on which the two floras almost meet, and when the greatest number of plants are in flower in all temperate climates,—the day on which, as if to become the favourite of the goddess of blooms, the great Linnæus was born.

As the spring advances the weather gets more and more variable. A cold, biting, north-east wind, with sudden sun-gleams casting a temporary lustre on the growing corn, a warm and drying day or two producing dust, rapid showers of snow or hail, not without bright sunshine, frosty nights or warm days,—alternate with each other, and succeed the rain, which often falls pretty copiously earlier in the season, so that the proverbial adages of " February fill-dyke," and of " March many-weathers," apply successively to the two portions of the spring.

One of the most striking phenomena of this season is the return of the vernal birds of passage, which arrive by degrees, and fill the woods and gardens with their song; those birds which remain with us all the year are now in full song; the Nightingale, the Redstart, the Blackcap, and the Willow Wrens arrive in April.

Among these birds of passage is the Swallow, which enjoys the distinction of having been mentioned by almost every writer on natural phenomena. The constant harbingers of spring, and living on winged insects that would be otherwise noxious to us, the Swallows have always been favourites. In Greece, the children used to make a sort of holiday on the first arrival of the Swallows. The Chimney Swallows appear between the 5th and 19th of April, and the Common Martlet and Sand Martin between the 20th and 30th of the month. There has been a great controversy as to whether the Swallows lie torpid in winter or migrate; but it is undoubted now that, like all other summer birds, they migrate to more southern countries in autumn.

Of course we must mention the mysterious Cuckoo, which now makes appearance, flitting from hedge to hedge, and from tree to tree, and—

> "Hid in some bush now sings her idle song
> Monotonous, yet sweet, now here, now there—
> Herself but rarely seen."

This bird must have some important function in the economy of nature, to be exempted from the toils of maternity, which it escapes by dropping its egg into the nest of some other bird, generally that of the Hedge Sparrow, who becomes the nurse of the young stranger. Why the Cuckoo cannot find time to attend to this tender business is indeed a mystery,—like that of many a human mother, who imitates the heartless bird, without the excuse of instinct or inability.

Insects are numerous in spring; on certain fine days thousands of species make their first appearance together. The early Sulphur Butterfly is soon followed by the Tor-

toiseshell, the Peacock, and lastly, the White Cabbage
Butterfly, towards the end of the season.

But the chief objects of the present season are the
march of vegetation in general, the development of the
leaves on the trees, and the flowering of plants. Now
Flora reigns luxuriantly; from the very commencement
to the end of the period, some new flower is added every
day, so that all we can do is to describe a few of the
most prominent plants of the season. In the meadows,
the first plant that covers them with a golden yellow is
the Dandelion, succeeded by the bulbous Crowfoot at
the end of April. Early in May, the creeping Crowfoot
in the uplands, and the Buttercups in the meadows,
cover the grass with the most brilliant golden yellow,
while in other places on shady slopes, and on ground
over which the trees may have been newly felled, the
field Hyacinth covers the entire surface with its rich
blue flowers. This mysterious springing up of plants on
ground cleared of timber is one of the many wonders of
vegetation. The Meadow Lychnis succeeds the Field
Hyacinth, and during the entire period of spring the
banks are covered with Primroses and Violets, and here
and there with Pilewort,—in the hedges, the Blackthorn first, and afterwards the Whitethorn, blossom.

In the orchard, a succession of blossoms on the Plum,
the Cherry, the Pear, and the Apple trees, give a remarkable richness to the face of nature, and convey the
idea of the rural goddess filling her horn of plenty.

In our gardens the richest variety of flowers, with
some of the most brilliant colours, gladdens the heart of
man. Some botanists have said that yellow is the prevailing colour both of spring and autumnal flowers
white of early spring, and red of summer flowers. This

however, is incorrect. We have all colours in nearly proportional quantity in each season. In spring the bright ultramarine blue of the *Cynoglossum Omphalodes*, or of the *Veronica Chamædrys*, which covers every bank in May, may be quoted as conspicuous examples of blue; the Harebell is also blue, and in some places is as common as the yellow Crowfoot. The early Van Thol Tulip (*Tulipa suaveolens*) and the Clarimond, *Tulipa præcox*, begin to blow out-of-doors early in April, amidst the remains of the early spring flora, and exhibit fine specimens of red and crimson. The Hyacinths of various colours, some of the Narcissi, and the Polyanthus of unnumbered dyes, all exhibit their flowers in the early part of this period, with Heartseases, and other flowers far from being yellow. Early in May the standard Tulips are in full blow, exhibiting every tint, stripe, and variety of colour. Toward the middle of the month, the rich crimson of the Peony may be compared with the bright light red of the Monkey Poppy.

Towards the close of the vernal season, the weather gets warmer, and is generally fine and dry, or else refreshed by showers; it is, however, seldom hotter than what may be called temperate, and the nights, when the wind is northerly, are still cold. Flora reigns triumphant; every hedge and bush and bank and field are in bloom. The blossom of the fruit trees gradually goes off, the grass in the meadows gets high, and partially obscures the yellow Ranunculus which decorated them in spring, and by the first week in June, the setting in of the solstitial season is manifest by the absence of dark nights and the blooming of a new set of plants.

With regard to its meteorology, spring commences on the 6th of March. Its duration is ninety-three days;

during which the mean temperature of London is raised, in round numbers, from 40 to 58 degrees.

The barometric mean is 29·93 inches.

The mean humidity is 78, saturation being taken at 100.

The mean temperature of the *season* is 48·94°,—the sun producing by his approach an advance of 11·18° upon the mean temperature of the winter. This increase is retarded in the fore-part of the spring, by the winds from *north* to *east*, which are then prevalent; and which form two-thirds of the complement of the season; but then the increase is proportionately accelerated afterwards by the southerly winds with which it terminates.

This period is in the first part characterized by a strong *evaporation* followed by showers, often with thunder and hail,—with which it is intimately connected in the latter part, as will be shown in the sequel.

The temperature of spring commonly rises, not by a steady increase from day to day, but by sudden starts —owing to the breaking in of sunshine upon previous cold, cloudy weather. At such times, the vapour appears to be, at times, thrown up in too great plenty, into the cold region above, where, being suddenly decomposed, the temperature falls back for a while, amidst wind, showers, and hail,—attended sometimes with frost at night.

Hailstorms are the attendants of spring, and Luke Howard describes one of the most remarkable, not only in its ravages, but also in its antecedent phenomena, which, duly considered, throws some light on the cause of thunderstorms in general. The thunderstorm in question occurred on the 19th of May, 1809, and is thus described by the great meteorologist himself:—

The day had been sultry, like some preceding ones, and overcast with clouds, which, during the afternoon, gave evident demonstrations of an approaching discharge of electricity. Large and deep *cumulo-strati* clouds were ranged side by side, mingled with *cirro-cumuli* and *cirro-strati,* the whole having that peculiar, almost indescribable character which these charged conductors assume, when wrought up to the highest state of electric tension.

" About five in the afternoon, being at the laboratory, and perceiving a continued roll of thunder, with vivid lightning approaching from the south, and the appearance of a heavy shower in that quarter, I anticipated a storm of no common violence. We were proceeding to take measures for the safety of some glass utensils, when in an instant there opened upon us a volley of hail of such tremendous force, as in ten, or at most fifteen, minutes, demolished most part of the skylights and south windows in the neighbourhood. These icy bullets, some of them a full inch in diameter, were discharged almost horizontally from a cloud to the windward, and in such quantity as to be drifted in large masses under the walls.

" Whether borne by the impetuous blast that came with them, or carrying the air thus before them, I could not determine, but such was the velocity of their motion, that in many instances a clear round hole was left in the glass they pierced; and one large pane had *two* such perforations distinctly formed,—the glass being otherwise entire and uninjured.

" The water in the river, lashed by the hail and raised by the wind, resembled a caldron boiling violently, rather than waves and breakers. The electrical discharges were incessant, approaching with the cloud, and passing off with it,—so that the whole resembled, in effect, the more mischievous artillery of human invention, —inspiring, in spite of philosophic reflection, and the delight at the grand and unusual phenomenon, a sense of actual danger.

" The sudden eruption over, it rained for a while moderately. The wind was at first E., then S. during the hail, then W., then E., then W. again. About seven, the clouds all at once put off their stormy character, and appeared as if going asleep after this prodigious expenditure of power. The remainder of the evening was calm and pleasant.

" The damage done by the hail was very great. A London newspaper estimated it, from the accounts which reached the editor, at 200,000 squares of glass broken in sashes, skylights, conservatories, hot-houses, etc., besides the injury done to the crops in fields and gardens. The foliage of large elms was cut off and scattered on the ground to a furlong's distance to leeward; and fruit-trees, besides being thus stript, received wounds in their bark which were visible long after. A West Indiaman, in passing Blackwall during the storm, had her fore and main-topsails blown over the side, and one man drowned."

Such is the vivid description of this thunderstorm as witnessed by Luke Howard; but elsewhere in his great book he makes certain observations which will here come in most appropriately. They relate to *evaporation*, which, as before said, seems to be intimately connected with electrical phenomena. The greatest evaporation which Luke Howard ever observed, occurred on the 17th of May, 1809, two days before the thunderstorm just described. On that day the amount of evaporation was 0·39 in.; on the following day it was only 0·28 in.; and on the next it fell 0·14 in.,—the corresponding mean temperatures being 67°, 70·5°, and 64°, and consequently furnishing, in respect of *heat*, no adequate cause for the decrease. It was on the evening of the day of *lowest evaporation* that occurred the tremendous thunderstorm before described, and thus it appears that *a rapid decrease of the daily evaporation in hot weather* may furnish a prognostic of approaching thunder and rain, when neither the barometer nor the hygrometer indicates the coming commotion.

As for the explanation of the cause, that suggested by Luke Howard seems perfectly in point:—the local influence of heat, aided by an electric charge in the air, had suddenly raised, as it were, a mound of vapour into those elevated regions of the atmosphere which it rarely visits in these latitudes, and where it is subject, from the contiguity of an intensely cold medium, to complete and extensive decomposition; in which seems to lie the true cause of the prodigious development of electricity manifested on these occasions.

We may further observe that the " suffocating heat," remarked before thunderstorms, is due to the want of evaporation on the surface of our bodies; it passes off as soon as the air is enabled by the outburst of the storm to favour evaporation.

There is another point connected with this subject, respecting which a few remarks may be acceptable,—we mean, the laws regulating the *paths* of thunderstorms. Nothing is, apparently, more capricious than the path or progress of a thunderstorm; but, after all, what else can be expected from a phenomenon which, powerful as it is, yields to the slightest attraction, and follows, so to speak, whithersoever almost anything may lead. Moreover, its discharges produce intense rarefactions, towards which denser and colder air must rush from any and every direction to restore the equilibrium : hence, doubtless, the variety of points whence the wind blows in thunderstorms. Hence, also, perhaps, the devious course it takes, for the winds it thus makes by its discharges must incessantly influence its path or progress. Hence, there may be places under its path, receiving neither hail nor rain, and others only a few drops of rain,—as in the thunderstorm of June 27th, 1866. The first storm seems to have formed between Windsor and Guildford, soon after one P.M., and moved slowly to the north, until it reached the Thames, near Egham, when it appears to have divided, one portion passing up the river to Windsor, and the other down it, over Hampton, Richmond, and Isleworth,—that is, in a north-easterly direction. At Gower House, Teddington, and at Richmond, the storm is reported to have come from the east, whilst half Richmond Park was dry and dusty, and an observer at Spring Grove, close by Richmond, had the brunt of the storm, but only registered 0·04 in. of rain, and reported the direction of the storm as S.S.W. to N.N.E.

It is only on the theory we have suggested that these perplexing results can be explained. The manifestations of the storm depend not only on its electric force but the effects which it produces, which, in their turn, influence the storm itself; and thus "gaps" occur in its wonted products or the injury they may inflict. Thus, in the storm in question, the largest hailstones did not do most mischief,—not wholly on account of their deficient

hardness, but simply because they were not impelled by
the rushing wind consequent to sudden rarefaction. So
they fell " quite gentle," as one of the natives told Mr.
Symons, at Willesden Green.

Such is the explanation we would suggest in a matter
as yet entirely undecided ; but still we deem it proper to
quote Mr. Symons, who offers the following hypothesis:—

" Some of our great authorities on wind have urged
the existence of vertical rotary motion to explain the
path of a thunderstorm, as if a cyclone, under ordinary
circumstances, be represented by a wheel revolving while
lying on its side, a *vertical* cyclone would be represented
by the same wheel revolving in its proper position on the
axle of a carriage wheel.   I have mentioned this because
it seems to me the readiest solution of a puzzling fact,
namely, that the largest hailstones did not do most mischief. Doubtless, this was partly due to the relative
hardness of the stones ; but it was not wholly this, as
many of the larger stones at Hampstead were of the
hard species, yet the damage there was trifling, as it was
beyond Cricklewood, although the hail was large and
and hard, they fell 'quite gentle,' as a native told me
at Willesden Green.   Now, supposing (for it is only a
bit of theory) there was such a vertical cyclone, its rotation being such that the motion of its lower half was in
the same direction as the general course of the storm, it
would solve at once several difficulties.   Thus, let its
lowest portion be assumed to have swept near the surface while crossing over Cricklewood, it would then give
wind-force sufficient to break the poplar, and by increasing the velocity of the hailstones would produce equal or
greater destructive power than where they fell without
the wind, and yet were larger.   Moreover, if this theory

be accepted, it will also explain how large and heavy hailstones could fall 'quite gentle-like' a mile or two further on; for, there, the upward motion of the lower advancing quadrant of the cyclone would, by its tendency to lift the stones, partly counteract the force of gravity and so let them fall 'quite gentle-like' to the earth."[1]

For our part, admiring the ingenuity of the theory, we cannot see the propriety of mixing the term "cyclone" with the subject,—the word having its precise meteorological significance, that cannot be applied to thunderstorms,—which, as before said, we take to be regulated both in their path and their effects by rarefactions and condensations; and the rush of air consequent to the former, constituting the various winds of the phenomenon.

On the whole, our island suffers but little from thunderstorms, compared with the fine fields of some provinces of France, which from time immemorial have been subject to their destructive visits. But human ingenuity, which has always been exercised in one way or another in an uncertain strife with the elements, has in France resorted to a bold and singular expedient in self-defence against the thunderstorm. The French actually blow up the nascent thunderstorm with gunpowder! The thing appears incredible, but it is nevertheless what Mr. Ruskin would call a downright fact.

This process, which is universal in that part of France called the *Mâconnais*, was originally introduced by the Marquis de Cheviers, a naval officer, retired on his estate at Vaurenard, who having recollected to have seen the explosion of great guns resorted to at sea in order to disperse stormy clouds, resolved to attempt a similar method

[1] Symons, 'Monthly Meteorological Magazine.'

to dissipate the hailstorms, whose ravages he had often witnessed.

For this purpose, he made use of boxes of gunpowder, which he caused to be fired from the heights on the approach of a storm. This had the happiest effect, and he continued till his death to preserve his lands from the ravages of hailstorms, while the neighbouring villages frequently experienced their baneful effects. He consumed annually between 200 lbs. and 300 lbs. of mining powder.

The inhabitants of the communes where the estate of the Marquis was situated, convinced of the excellence of the practice, from the experience of a great number of years, continued to employ it. The size of the powder-boxes, their charge, and the number of times they fire them off, vary according to circumstances and the position of the places. In the commune of Fleury they use a mortar which carries a pound of powder at a charge; and it is generally upon the heights, and before the clouds have had time to accumulate, that they make the explosions, which they continue until the storm-clouds are entirely dispersed and dissipated. The annual consumption of gunpowder for this purpose is from 1300 lbs. to 1600 lbs.

Such is the French mode of getting rid of these troublesome celestial visitants; but in thus preventing partial damage to their crops, it should be considered whether the fertilizing rain may not also, for the time, be driven from the district.[1]

[1] If the digression be allowed, we may state that this is not the only way in which man has contrived to counteract the elements. It is stated by Boussingault that in the plains of Cusco, in Peru, on still clear evenings, the Indians set fire to a heap of wet straw or dung, and, by this means, raise a cloud of smoke which prevents frost. This is

But if it be scarcely advisable to resort to any means to interfere with thunderstorms, the want of skilful observation of their gathering, and the probable cause of the waters which they may discharge, may be attended with the most disastrous results, especially in a hilly district, as occurred in 1807 at Silkstone, in Yorkshire. The clouds had portended rain, but none had *fallen there*, when suddenly a torrent of water deluged the town, which is situated in a valley, and several persons were drowned. The greatest transition from cold to heat ever remembered *had been observed* during the week, but otherwise unheeded, and this inundation was caused by a mass of clouds, during the thunderstorm, bursting in a field at Bradfield.

The heat and vapour of our spring, notwithstanding the interruptions caused by thunderstorms, accumulate on the whole; and the atmosphere now receives an addition of 0·30 in. upon the mean of the winter.

The *barometer* averages 29·83 in.; but the extreme elevations and depressions of the column go off, in a great measure, during the season; and by the end of spring, the range is contracted to about an inch and a half. The mean range of the season is 1·81 in.

The *evaporation* amounts to 8·85 in., being about a sixth part *more* than the proportion which would correspond to the mean temperature.

quite in accordance with the teaching of science. Smoke and dry fogs, by obstructing the radiation of caloric, or heat, into free space, enable low grounds to retain their warmth during the night. Curiously enough, the peasants of Chamouni, in Switzerland, do the same precisely, if their crops should not have ripened towards the end of the season,—burning greenwood on the two sides of the enclosing mountains of their valley, the smoke of which, uniting in the middle, forms a kind of cloudy canopy, which is found not only to prevent the escape of radiated heat, but also to increase it, and to prevent frost.

Consistently with this proof of *dryness*, the average humidity indicated by the hygrometer is 60.

The mean *rainfall* is 4·81 in. It increases at a small rate through the season; but being greatly exceeded by the evaporation, the soil uniformly gets dried; and the small springs, which issued during the winter from the superficial strata, disappear or become insignificant.

The lower atmosphere becomes very transparent in the forepart of the season; but the brilliancy of the returning sun is apt to be dimmed or eclipsed during pretty long intervals, by a close veil of cumulo-stratus clouds, which cover the whole sky with their drapery, connected at certain points by a kind of central stem, or basis of the structure, hanging low in the sky. At other times, under the course of easterly or northerly winds, there appear regular ranks of a meagre cumulus, coming on from the horizon, and passing away to the opposite quarter, with little or no change of form or magnitude, and unattended in great measure by any other modification of cloud. But in the latter part of the season we have a variety of clouds.

The *cirrus*, which is connected with variable breezes throughout the year, now assumes more tint and consistency, and is particularly fine before thunderstorms.

Majestic *nimbi* traverse the sky in succession, affording slight showers of large opaque hail or snow, the prodigious electricity attending which seems to prove that these singular clouds really act as *conductors*, fitted by communicating a portion of the repulsive fluid, to prepare the way for the descent of subsequent showers, without the necessity of an explosion. In this case it appears that the centre of the shower is *positively* charged, while the circumference is *negative,*—a fact which affords a clue for explaining many of the most sudden, and apparently capricious, changes discoverable by the insulating-rod, when showers are flying about in *distinct bodies,* the separate charges of which must, independent of their own composition, produce many phenomena by influencing each other.

It is remarkable that a *snowstorm,* in the middle of spring, not unfrequently proves to be the forerunner of *the first hot weather,* which is developed *in ten days, or at most two weeks after it.*

Consistently with this fact, some of the swallow tribe, of which several species come from the south to avail themselves of our temperate summers for breeding, if not also to shun the tropical rains, make their first appearance in the midst of such weather. This seems to prove that their approach is not gradual, but rather a rapid flight to our shores by the help of a superior *southerly current;* and some observations on the phenomena consequent on their disappearance, induced Luke Howard to suspect that they avail themselves of similar aid from a high northerly current to return.

A wet spring seems not at all ungenial in our climate, provided it be followed by a warm and dry summer; but in general, dry weather, however cold in the *early part* of the season, appears to be the wish of our farmers, who have no objection to showers after they have got their seed into the ground. But should the farmer have too much rain for the business in which he may be occupied, it may be some consolation to him to know that the circumstance indicates a dry time for the ensuing harvest.

III.—SUMMER.

"From brightening fields of ether fair disclosed,
Child of the sun, refulgent Summer comes,
In pride of youth, and felt through Nature's depth.
He comes attended by the sultry hours
And ever-fanning breezes on his way."

Early summer, or the solstitial season just alluded to, is perhaps the most delightful of the whole year; for, although the spring is the most adorned with blossoms, yet the days have now attained their full length, a beautiful twilight takes the place of night, and we seldom or never feel cold, except in particularly unseasonable years. Besides, the air is generally calm, without storms, and wholesome; and though sometimes great heat prevails, yet it is relieved by thunder-showers, and the evenings are refreshing and delightful.

The approach of the summer solstice is indicated by full-grown grass in the meadows, the flowering of the Purple Clover, the Midsummer Daisy, the Yellow Rattle (*Rhinanthus Crista-galli*), and in the cornfields, the Red Poppy. In our gardens the Scarlet Lightning (*Lychnis Chalcedonica*), the Sweet-William (*Dianthus barbatus*), Pinks, and the whole of the beautiful tribe of Roses—the queen of flowers—besides numerous other plants, are peculiar to this season, and would be a certain mark of its presence to any botanist who might, after a long voyage, be shipwrecked without any almanac, on our shores.

Sheep-shearing takes place early in this season. The flowering of the Elder is a phenomenon of the early part; the Hawthorn still continues in bloom, but the fruit-trees are out of flower, and the fruit is " set "

—the period anxiously looked forward to by the husbandman, respecting the orchard, for it is on the " setting" of the flowers that the abundance of the autumn
yield depends.[1]

Most of the Lilies flower in this season ; the Yellow
Pompon is the first, the Orange Lily follows, last the
Turk's-cap and the White Lily.   In the early part of
summer the major part of the species of Iris bloom.

Some fruits are ripe towards the end of the season ;
scarlet Strawberries come in season about the 15th of
June, the larger sorts before Midsummer Day; Maydock
Cherries ripen at the same time, and the first week of
July generally colours the red, white, and black Cur-
rants.   When the summer fruits are backward, there is
always danger of a late and wet harvest.   Haymaking is
the characteristic of early summer ; in the immediate
neighbourhood of London it is usually over a week or
ten days sooner than in the country.   This part of the
season often closes with very hot weather, which gives
place to the summer rains during the latter part.   The
last fourteen days are called the Dog Days.

The hottest period begins about St. Swithin's Day,
July 15th, and continues till Michaelmas.   It is on the
whole the hottest period of the year, but the heat gradually declines, and towards the close of the season the
nights begin to get cold, and the daily temperature is
much diminished.

One remarkable circumstance of this season is the
silence of the grove,—nearly every bird ceasing to sing,
and continuing mute till near the close of the season,
when they begin to sing a little again.

[1] " Quotque in flore novo pomis se fertilis arbor
       Induerat, totidem autumno matura tenebit."

Many birds now begin by degrees to congregate, and form large flocks ranging the cornfields together. Starlings are flocked by the end of July, and linnets by the middle of September; the swifts leave us about the 15th of September, which is the festival of the Assumption in the Catholic calendar,—and nothing but an accidental straggler is to be seen left behind.

The summer flowers cannot be called the most beautiful of the year, though if well managed a great display of colours may be produced in the garden. The Dahlias, —once so fashionable,—China Asters, French and African Marigolds,—prettier than they smell,—Chrysanthemums,—the gigantic Sunflower,—emblem of true love,—and a great variety of other syngenesious plants flower during this period, and many of them continue till late in the autumn. In the fields the flowering of the Yellow Autumnal Dandelion (*Aspargia autumnalis*) gives to certain meadows the appearance of a second spring.

Mushrooms and a few other fungi appear towards the close of this period, and are particularly abundant when, after a dry summer, the summer rains set in late and abundant. The cultivation of mushrooms is said to be one of the most profitable occupations.

Of course the golden corn, the appearance of a ready harvest, and everything in seed, tell of mature summer.

The season closes with a considerable reduction of temperature, and a diminished evaporation, producing mists, fogs, and a moist atmosphere in general.

With respect to its meteorology, summer begins with the 7th of June and lasts ninety-three days.

The mean temperature of the season is 60·66°, or 11·72° above the spring. The mean of the twenty-four hours rises during the season from 58° to 65°, but returns before the close to the former level.

The mean height of the barometer for summer is 29·87 in., or 0·04 in. above the vernal mean. The degree of humidity is 77, saturation being taken at 100.

The atmosphere now acquires, under the vertical rays of the sun, now in full north declination, the greatest quantity of heat and vapour which it at any time contains; and it weighs most by the barometer. The range of this instrument still diminishes to the middle of the season, when it does not exceed an inch; it then gradually increases again to the end. The mean range is 1·08 in.

The great fluctuations in the density of the atmosphere are principally due to our participation, by turns, in the polar and equatorial or tropical atmospheres, between which we are situated. But our position in summer when by the inclination of our pole towards the sun we are presented in a more direct manner to the rays of the sun, approximates our climate to those of the equatorial regions. Thus we become more uniform both in temperature and density than in any other season— though still much more variable in both respects than the countries in that part of the globe.

It is this approximate equalization of our atmosphere in density and temperature to that of the equatorial regions, and the increased temperature of the more northern latitudes as well as polar regions, which render our summer generally a season of calms, or at any rate prevent the occurrence of such storms as occur at other seasons. A storm presupposes a rush of air from one region towards another, attracted either by difference of temperature or density. Now, from the causes just mentioned, this cannot take place, since something like a general equilibrium exists; hence the comparative quietude of summer in general. The barometer, therefore, varies much less in summer than in winter, when opposite circumstances exist; but its movements in ascent and descent in summer are not, therefore, less indicative of those changes in the density of the air on which the weather, in a very considerable degree, depends.

In most districts haymaking is chiefly conducted within the limits of this season. This branch of rural economy has derived very considerable aid from the use of the barometer; indeed we may say that much less valuable fodder is spoiled by the wet at present than in the days of our forefathers. Still there is much room for improvement in the knowledge of our farmers on the subject of the atmosphere, which concerns them so nearly. Instead of trusting to weather-prophets, why not take the trouble to get the chief facts of the subject and apply their principles to all occasions, which is no difficult matter? From the records of the past and existing occurrences it is always within our power, at the beginning of our summer, to know with more or less certainty whether the season will be a wet or a dry one; or to discover, in a certain period of it, that the average quantity of rain to be expected for the month has already fallen.

In this case, if there have been no signs of a wet year, it is obvious that the farmer may proceed with greater confidence in his operations; if, on the contrary, the year, from the rainfall of the first six months and the weather of May (as explained in Chap. XIV.) must be a

wet one, then the utmost vigilance will be required to
secure the fruits of the harvest. When, therefore, there
is reason from this kind of information to expect much
rain, the man who has courage to begin his operations
under an unfavourable sky, but with good ground to
conclude, from the state of his weather-instruments and
his knowledge of weather principles, that a fair interval
is approaching, may often be profiting by his observations; while his cautious neighbour, who has waited
"for the weather to settle," may find that he has let the
opportunity go by. Now, this superiority is attainable
by a very moderate share of application to the subject,
and by keeping a simple diary of the barometer and the
rain-gauge, with that of the all-important hygrometer
and the wind-gauge or vane under his daily notice. All
seasons may vary, but none of them will be found incon-
sistent to themselves, simply because there is a reason
for every atmospheric phenomenon, and there is very
little difficulty in discovering it if we look carefully into
the subject. Temperature, moisture, pressure—these are
the three keys that open the secret boxes of the weather
in all seasons and on all occasions.

The prevailing winds of our summer are clearly
W.–N., or those which range from west towards north,
the latter point not included.

The mean evaporation is 11·58 in., being above a
fourth part more than the proportion which should correspond to the temperature.

The humidity is 52 throughout the season.

The rainfall is 6·68 in. Our summer rains, which
are much the most plentiful in the middle of the season,
or during July in general, appear to be the result of a
less powerful and less constant operation of the causes

which produce the rains of the tropics. Hence, our wettest summers are those in which, by the concurring effects of the sun's declination and the currents, we partake the most of the *tropical* atmosphere; and we get a dry summer only by approximating, in consequence of an opposite course of winds, or an atmosphere generally calm or breezy, to the circumstances of the high northern latitudes. In the one case our summer is, so to speak, equatorial; in the other, polar. In the one case, we seem to be placed in the great general stream of subsiding air from the south; in the other, the air returning from the north, after having deposited its excess of water.

Therefore, a North-West current is our *fair-weather wind.*

This will bring us moderate weather and sunshine, *so long as it is not interfered with by southerly currents;* and these, as we have shown, arrive, with rain and thunder, from an *easterly* direction,—consequently in a way most likely to mix with, and be decomposed by, the prevailing westerly current. (See *ante,* p. 34.)

In this state of things, or when there is a tendency to this process, our summer clouds exhibit a magnificence approaching to that of the tropical sky. This results from the greater quantity and more elevated situation of the vapour in the atmosphere.

The *cirrus,* which is usually the first to make its appearance after serene hot weather, now spreads its tufts to a greater extent, and assumes a more dense appearance than during spring; and the *cumulus*—ever beautiful and of favourable aspect when insulated in the midst of sunshine—now tends constantly to inosculate or unite above, or to become grouped laterally, with the *cirro-*

cumulus and *cirro-stratus*, which occupy the middle region.

From the mixture of these, and the interspersion of a quantity of *anomalous haze*, in patches or extensive beds, there results a sky more readily remembered than described,—and which is very easily and suddenly resolved into thunder-showers.

The locality of these is often determined by the situation of a rapidly growing *cumulus*, which, becoming a centre of union for the surrounding looser portions, gradually extends itself above and around, until it has put on the form of *cumulo-stratus*—the last stage before the explosion which decides the precipitation of the water in heavy rain. This once begun, the *nimbus*, with a confused moving and spreading sheet, increasing the obscurity on all sides, renders further observation from below very imperfect. At every interval, however, of some hours' duration with the same winds—the same state of the sky returns again.

A tendency to *rain* in such a sky is perhaps as decidedly indicated by the grouping of the *cirrus* and haze in certain parts, in the form of a crown of a *nimbus*, as by any other observed symptom; whilst the cirro-cumulus, which is the proper natural index of a rising temperature, is favourable to *dryness* and fair weather, except when it forms a part of the preparatory machinery of thunderstorms. In the latter case it is usually arranged on a kind of arched base, mixed with the *cirrus* and *cirro-stratus*, and the whole with the haze above-mentioned. The immediate tendency to an electrical explosion is always indicated to those who have the view of the lower part of the cloud, by a surprisingly quick motion of the loose ragged portions of *scud* around it,

which seem in haste to obey the powerful attraction of
the mass, and take their places in the general arrangement, on which, probably, the effect depends.

A thunderstorm in profile on the horizon, in the dusk
of the evening, is one of the most sublime spectacles in
nature.   Such is the immense depth and extent, and
the picturesque forms and complex arrangements of
these natural batteries of " heaven's artillery" before
the explosions ; and when these have commenced, it is
easy, for a while, to discover the very cloud from which
each discharge proceeds, the whole substance of it becoming, at the moment, *incandescent* with electric light—
like the explosion of a magazine.   In proportion as the
charge expends itself, the high-wrought forms of the
clouds disappear, the crowns of the *nimbi* are spread
out above, and the whole passes into the familiar appearance of a distant bank of showers, which, in effect,
is now the general consequence.

Often in summer, with a northerly and easterly
wind, the skies will be densely clouded, and showers
will be expected,—but no rain falls.   The *hygrometer* is
our safe guide on such occasions ; its certain indications
will permit the farmer to proceed fearlessly with his
labours.   High above that mass of cloud the sun is
shining brightly, as is evident by the occasional gleams
of light penetrating to the earth through the breaks.
Thus placed between the radiating earth and the glowing fire of the sun, the dense mass of cloudy vapour is
held in suspense, until other conditions of wind, namely
from the west, shall decompose and precipitate it in
showers.

IV.—AUTUMN.

" But see the fading many-coloured woods,
  Shade deepening over shade, the country round
  Imbrown, a crowded umbrage, dusk and dun,
  Of every hue, from wan declining green
  To sooty dark."

The main characteristics of Autumn are departure and decay,—the sensible departure of the sun and diminishing sunshine,—the departure of the winged minstrels that enlivened the summer,—the decay of the leaves, changing colour from green to yellow, red, or brown, and at length, falling, in the silence of the calm night, or tossed by the furious wind when the storm rages. Hence the old name *Fall*, given to Autumn, and still generally heard from the lips of our cousins and descendants in the United States of America. The Beech, the Oak, and a few deciduous trees keep their old leaves till the spring.

The many-coloured leaves of autumn contrast with the green grass and the blue skies and the soil, and make a picture which we find pleasure in beholding; whilst the passing away of the bright summer and the coming of wintry gloom which impends, suggest solemn thoughts which increase in intensity as our years roll on—as our hopes fail—as the friends of our youth and those whom we love leave us, leave us—but not for ever !

The retreat of the Swallows and the Martlets constitutes one of the most remarkable features in the history of the season. Swallows assemble early in September, and so continue to appear in vast numbers, roosting on

the tops of houses and lofty buildings, and the greatest
part of the species migrate between new and old Michaelmas Day.   Martlets retire a few days later; straggling
Swallows are seen about till the end of that month.

On the other hand, many other birds now arrive in
flocks; Wild Geese and Ducks perform partial migra-
tions, and Woodcocks and Snipes arrive.

The only specific flowers of autumn are Saffrons, the
autumnal Crocus, the purple and the white varieties of
the Colchicum in our gardens, together with Michaelmas
Daisies, the other late Asters and Chrysanthemums, the
gorgeous varieties of which strive to keep up a semblance
of floral beauty at a time when cold air, wet fogs, and
alternations of wind and fine weather induce us to look
elsewhere for the pleasures of the senses.

There is something in the final appearance of this
season, just as we step from Autumn into Winter, that is
well calculated at all times to excite feelings of a melancholy interest.   The garden is rife with homilies on the
" wreck of matter," and mementoes of mortality are
abundantly depicted in the withered flower and droop-
ing shrub ; whilst the woodlands and groves are no less
prolific in memorials of all things passing to their original "dust."   There is a funereal characteristic about
Autumn which none of the other seasons possess ; she
is the messenger of fruition and death.   Fruits ripen
and flowers wither at her approach, and this power does
not cease until vegetable vitality is subdued and laid
prostrate.   Spring is the season of hope ; the first crocus
that peeps from beneath its pure white mantle of snow
is greeted with gladness, because it is the precursor of
brighter and more beautiful flowers.   Summer is the
season of birds, blossoms, and fruit; Nature then puts

on her richest jewellery, and we are dazzled by the splendour and the beauty of their colours. Autumn is the season of plentitude—but then, before we have scarcely done gazing at the lovely products of Pomona, the " sere and yellow leaf" parts from its spray, and rustling scarcely audibly along, rests at our feet, warning us to prepare for another change. It is not, however, amidst the fogs and smoke of a city atmosphere that this change can be felt. Autumn, to be appreciated, must be enjoyed some miles from town. The " green and yellow melancholy" which there steals over the landscape, and the mild and steady serenity of the weather, with the transparent purity of the air, speak not only to the senses, but to the heart. There is a silence in which we hear everything,—a beauty that must be observed. The *cinquefoil*, with one lingering blossom, yet appears, and we mark it for its loveliness.

Rambling with unfettered grace, the tendrils of the Bryony festoon with its brilliant berries the slender sprigs of the Hazel and the Thorn ; it adorns their plainness, and receives a support which its own feebleness denies.

The Agaric, with all its hues, its shades, its elegant variety of forms, expands its cone, sprinkled with the freshness of the morning,—a transient fair, a child of decay, that was born in a night, and will perish in a night.

Anon, the Jay springs up, and screaming tells of danger to her brood. Then comes the loud laugh of the Woodpecker, joyous and vacant—like a madman ; the hammering of the Nut-hatch, cleaving its prize in the chink of some dry bough ; whilst the Humble-Bee, torpid in the disk of the purple thistle, just lifts a limb to pray forbearance of injury,—to ask for peace, and bid us—

" Leave him, leave him to repose !"

All these are distinctive characteristics of the season, marked in the silence and sobriety of the hour, and have left, perhaps, a deeper impression on the mind than any afforded by the profuse luxuriance of summer, or the verdant promises of spring.

With respect to its meteorology,—

Autumn begins on the 8th of September, and lasts 90 days. Its mean temperature is 49·37°, or 11·29° below that of the summer; the daily range is 18°.

The mean height of the barometer is 29.78 in., being 0·09 in. lower than that of the summer. The range increases rapidly during this season, the mean extent being 1·49 in.

The prevailing winds are the class S.–W. throughout the season.

The evaporation is 6·44 in., or a sixth part less than the proportion that would correspond to the temperature.

The humidity is 84 degrees, saturation being taken at 100.

The mean rainfall is 7·44 in.; the proportion of rain increases from the beginning to near the end of the season. This is the true rainy season with us.

The earth, which has become dry to a considerable depth during the spring and summer, now receives again the moisture required for springs and for the more deeply-rooted plants, in the following year. Autumn ministers to autumn from year to year, in each case provident of the ensuing harvest.

In their meteorological aspect, the changes in the state of the atmosphere in autumn are still referable to one and the same cause—the return of the sun to the south.

The heat declining daily, the store of vapour in the atmosphere undergoes a continued decomposition,—the loss of weight arising from this decomposition not being made up, as in summer, by an equal production of new vapour. Hence, a declining barometer, with extensive rains, chiefly in the latter part of the season. Thus, the mere presence of vapour is not a means of diminishing pressure, without an increase of temperature. The whole increase of pressure, derived on the average of the barometer in spring and summer, is thus disposed of, and the atmosphere returns to its minimum weight.

From the more saturated state of the air, the evaporation falls short of the temperature, and the hygrometer, at the same temperature, exhibits an average of about 8 degrees more moisture than that of the spring.

If we regard only the sky, the fore part of the autumn is the most delightful of the year in our climate; and in its genial influences we readily forget or fail to notice the serious meteorological changes of the season. Who can feel otherwise than comfortable, when the decomposition of vapour, from the decline of the heat, is as yet but in its commencement, or while the electricity remaining in the air continues to give buoyancy to the suspended aqueous particles,—a delicious calm often prevailing for many days in succession, amidst perfect sunshine, mellowed by the vaporous air, and diffusing a rich golden tint as the day declines upon the landscape! Then it is, chiefly, that the *stratus* cloud, the lowest and most singular modification of cloud-land, comes forth in the evenings, to float over the low plains and the valleys, and shroud the landscape in a veil of mist until re-visited by the sun at dawn. So perfectly does this inundation of suspended aqueous particles imitate

R

real water, when viewed in the distance at break of day, that the country people themselves are deceived by its unexpected appearance, taking it for streams of the crystal fluid.

A phenomenon attends this state of the air, too remarkable to be passed over in silence, especially as it is connected with the weather. An immense swarm of small spiders take advantage of the moisture of the season to carry on their operations, in which they are so industrious that the whole country is soon covered with the fruit of their labours in the form of a fine network, which is called *Gossamer*. They appear exceedingly active in pursuit of small insects, which the cold of the night now drives down, and commence this fishery about the time that the swallows give it up, and quit our shores. Their manner of locomotion is curious; half volant, half aeronaut, the little creature darts from the papillæ on his posterior extremity a number of fine threads which float in the air. Mounted thus in the breeze, he glides off with a quick motion of the legs, which seem to serve the purpose of wings for moving in any particular direction. As these spiders rise to a considerable height, in very fine weather, and when it is likely to last, their tangled webs may be seen descending from the air in quick succession, like small flakes of cotton. On threads of stronger texture, produced by some of these autumnal spiders, and which are often extended from tree to tree, or across country roads, frequently broken by the face of the passer-by, and some yards in length, the most minute dew-drops collect in close arrangement; and on the first touch of the support, run together and fall down,—thus giving a practical illustration of the manner of *the formation of rain in the atmosphere!* For, it is by the union of infinitely small globules of condensed vapour that the drops are formed which ultimately fall as rain, increasing in size according to the height from which they descend.

Both on these gossamers and on webs placed in an oblique or vertical direction, on the shrubs and herbage, and formed with the symmetry usually displayed by this insect, these drops are occasionally found *frozen*, and thus a string of little ice-beads may be taken up with the web. Nor is that all;—from the texture thus covered with dew, the sun's rays at times reflect innumerable little rainbows. Such are the smaller beauties of the autumnal landscape, and we need not be fairies to appreciate them.[1]

Autumn is remarkable for its heavy dews, owing to the depression of the temperature during the nights. These are sometimes so abundant as to admit of measurement in the rain-gauge. On one night, towards the end of September, Luke Howard got one-hundredth of an inch of water in one night from dew, and in the last six days of October eleven-hundredths from copious dews and mists.

In the drops of dew, when of considerable size, and under the clear morning sun, the meteorologist may observe a good instance of the reflection and refraction which produce the rainbow. He has only to place himself with his back to the sun, and singling out a particular drop which appears brilliant with any colour,

---

[1] " In crossing the Channel from Calais to Dover," says Forster, " I have observed that the captains of the vessels have sometimes prognosticated fine settled weather from the settling on the masts and rigging of a certain sort of web, which we take to be the woof of some spider though we have observed it to alight on the ships when some way out at sea."

he may, by changing his position so as to vary the angle, and keeping his eye on the drop, draw out the different prismatic tints in succession.

Of a totally opposite character are the fogs of autumn. When fog becomes visible, it is because the air is saturated with moisture; then only can the vapour of water be incessantly precipitated for several hours. It is important to insist on this circumstance; for Deluc and others, who employed imperfect hygrometers, have affirmed that the air is often *very dry* in regions where fogs are forming. The experiments of De Saussure prove the contrary. If the hygrometer be suspended before a window in the centre of a city, undoubtedly it cannot indicate the degree of saturation during the times of fog; but this occurs because the instrument is warmed by the walls of the building; even this anomaly eventually disappears when the fog remains for several hours.

The formation of fog is often very different from that of dew. When dew is deposited, the soil is always colder than the air; when fog occurs, the contrary is the case. The moist soil is warmer than the air, and the vapours that ascend become visible, like those which rise from boiling water, or like the vapour of expired air, which, in winter, condenses the moment it escapes from the mouth. So in autumn, we frequently see fogs above rivers, the water of which is much warmer than the air before sunrise.

In countries where the soil is moist and warm, and the air moist and cold, thick fogs may be expected. This is the case in England, the coasts of which are washed by a sea at an elevated temperature.[1]

[1] Knemst, 'Meteorology.'

London fogs have become proverbial. One of the most remarkable was that which occurred on the 12th of November, 1828. It began to thicken very much about half-past twelve o'clock, from which time, till near two, the effect was most distressing, making the eyes smart, and almost suffocating those who were in the street, particularly asthmatic persons. In the City, all the bankers' and offices of different description, as well as the principal shops, were obliged to have lights. To see with any distinctness further than across the street was impossible; all the narrow lanes, beyond the perspective of a few yards, were absolutely in a state of darkness; and in the great thoroughfares, the hallouing of coach-men and drivers to avoid each other, seemingly issuing from the opaque mass in which they were enveloped, was calculated to awaken all the caution of riders, as well as of pedestrians who had to cross the streets. On the Thames, as on land, the tendency which fog has to enlarge distant objects, was strikingly illustrated; the smallest vessels on their approach seemed magnified to thrice their dimensions. St. Paul's had a prodigious effect through the mist, though neither that nor the Monument were visible above the height of the houses. This optical illusion results from the fog diminishing the brightness of objects, and consequently suggesting a greater distance; since, while the visual angle remains the same, the greater the distance the greater the apparent magnitude. This fog cleared off a little about a quarter past two, but returned with all its density in the evening.[1]

As before explained (p. 69), smoke from coals plays a notable part in the fogs of large towns and cities.

[1] 'Public Ledger,' November 14, 1828.

Dense fogs in autumn and winter are generally followed by hoar-frost and then by wind and rain.

We come now to the consideration of somewhat more extensive matters—the gales of autumn. The latter part of this season and the beginning of winter are more peculiarly subject to gales of wind from the south-west.

While our north-east breezes are plainly the result of sunshine, and blow almost exclusively by day, the south-west winds appear to prevail chiefly by night,—the one forming part of an ascending, the other of a descending set, of currents. That our westerly gales come from above, is manifest from the manner in which the clouds indicate, beforehand, the increase and decrease of velocity which they afterwards manifest below. Their violence is well known, and there seems no way of accounting for their occasional excessive force, but by attributing it to the westerly *momentum* which the air acquires in a lower latitude, by revolving in a larger circle about the earth's axis, and which it may bring with it when suddenly translated northward. For, the portion of the earth whence it issues rotates at a much greater velocity than those against which it impinges in our latitudes, so that its proper speed is added to that with which it is drawn towards our colder latitudes by the difference of temperature.

It appears, moreover, that the sudden depressions of the barometer, of a few hours' duration only, which accompany these gales, and exhibit their minimum point about the time of the greatest force of the wind, are due to *an actual loss of gravity by the centrifugal force of the air, for the time.* Indeed, it is owing to the fact that the equatorial or heated current moves with its original equatorial speed or thereabouts, and therefore *faster* than the northern regions when it arrives there, maintains its *westerly* direction. Just the contrary happens to the polar current, which, as it advances towards the equator, meets with portions of the earth endowed with greater and greater velocity, so that it is, as it were, left behind, and thus produces the effect of a current moving in the opposite direction, that is, *eastward*, instead of north, its original direction, and not westward, with the earth itself.

The tremendous hurricanes of the tropics have sometimes been closely approached in violence by our autumn gales; and a very graphic account of one of these in the olden time may perhaps form a fitting conclusion to this part of our subject. The description is in an old book, of 272 pages 12mo, entitled 'THE STORM, or a Collection of the most remarkable Casualties and Distresses which happened in the late dreadful Tempest both by Sea and Land. London, 1704.' The motto of the book is most appropriate:—"The Lord hath his way in the whirlwind and in the storm, and the clouds are the dust of his feet." (Nah. i. 3.)

The date of this terrific tempest, as to its extreme violence, is the night of the 26th–27th of November, old style, 1703, being about the time of the new moon. It appears to have been preceded by a very wet season for about six months.

"It had blown so exceeding hard for about fourteen days past that we thought it terrible weather; several stacks of chimneys were blown down, and several ships lost, and the tiles in many places blown off the houses; and the nearer it came to the fatal 26th of November the tempestuousness of the weather increased.

"On the Wednesday morning, being the 24th of No-

vember, it was fair weather and blew hard; but not so
as to give any apprehensions, till about four o'clock in
the afternoon, the wind increased, and with squalls of
rain and terrible gusts blew very furiously. The wind
continued with unusual violence all the next day and
night; and had not the *Great Storm* followed so soon,
this had passed for a great wind.

"On Friday morning it continued to blow exceeding
hard, but not so that it gave any apprehensions of danger
without doors. Towards night it increased; and about
ten o'clock our barometers informed us that the night
would be very tempestuous—the mercury sank lower
than ever I had observed it on any occasion. It did not
blow so till twelve at night, but that most families went
to bed; but about one, or at least by two o'clock, 'tis
supposed, few people that were capable of any sense of
danger, were so hardy as to lie in bed. And the fury
of the tempest increased to such a degree, that as the
editor of this account, being in London, and conversing
with the people the next day, understood, most people
expected the fall of their houses. And yet, in this
general apprehension, nobody durst quit their tottering
habitations; for whatever the danger was within doors,
'twas worse without. The bricks, tiles, and stones,
from the tops of the houses, flew with such force, and
so thick in the streets, that no one thought fit to
venture out, though their houses were near demolished
within.

"It is the received opinion of abundance of people,
that they felt, during the impetuous fury of the wind,
several movements of the earth; and we have several
letters which affirm it. . . . And yet, though I cannot
remember to have heard it thunder, or heard of any

that did, in or near London, in the country the air was
seen full of *meteors and vaporous fires;* and in some
places both thunderings and universal flashes of lightning, to the great terror of the inhabitants.

" From two of the clock the storm continued, and
increased till five in the morning; and from five to half
an hour after six, it blew with the greatest violence.
The fury of it was so exceeding great for that particular
hour and a half, that if it had not abated as it did, nothing could have stood its violence much longer. In
this last part of the time, the greatest part of the damage was done. Several ships that rode it out till now,
gave up all—for no anchor could hold. Even the ships
in the Thames were all blown away from their moorings,
and from Execution Dock to Limehouse Hole there were
but four ships that rode it out; the rest were all driven
down into the *Bight,* as the sailors call it, from Bell
Wharf to Limehouse, where they were huddled together
and drove on shore, heads and sterns one upon another,
in such a manner as any one would have thought it had
been impossible. . . .

" This sort of weather held all Sabbath-day, and Monday, till on Tuesday afternoon it increased again, and
all night blew with such fury that many families were
afraid to go to bed. At this rate it held blowing till
Wednesday about one o'clock in the afternoon, which
was that day sevennight on which it began—so that it
might be called one continued storm from Wednesday
noon to Wednesday noon. *In all which time there was
not one interval in which a sailor would not have acknow-
ledged it blew a storm;* and in that time two such terrible nights as I have described."

Such a tempest could not be supposed to be limited

to this island; accordingly it appears to have spread over a great part of the north of Europe, though nowhere with equal impetuosity as with us. Over most parts of South Britain and Wales, the tallest and stoutest timber-trees were uprooted, or snapped off in the middle. It was computed that there were twenty-five parks in the several counties, which lost a thousand trees each,—the New Forest, Hants, above four thousand. Whole sheets of lead were blown away from the roofs of strong buildings; seven steeples, above four hundred windmills, and eight hundred dwelling-houses, blown down; and barns, out-houses, and ricks in proportion, besides a great destruction of orchards. About one hundred and twenty persons lost their lives on land,—among whom were the Bishop of Bath and Wells and his lady, who unhappily lodged in a ruinous castle,—also the engineer who had erected the then lighthouse at the Eddystone, who was blown into the sea along with the structure, which he had promised himself would bid defiance to the elements.

At sea, there were few ships to sink—the previous terrible weather having brought them into port in very unusual numbers—but in the harbours and roadsteads of England, so many vessels ran foul of each other and sank, or foundered at anchor, or were driven on the sands, or to sea where they were never heard of, that it is computed eight thousand seamen at least perished on the occasion. A vessel laden with tin, being left in the small port of Helford, near Falmouth, with only a man and two boys on board, drove from her four anchors at midnight and going to sea, made such speed before the wind, almost without a sail, that at eight in the morning, by the presence of mind of one of the boys, she was put into a narrow creek in the Isle of Wight, and the crew and cargo saved.

This run may give us some conception of the velocity of the wind; for, if we consider that the course of the vessel, even by the winds, could not have been direct, but in a large curve outwards from the coast, the rate at which she went exceeded thirty miles an hour on the average, and that of the wind must have been three or four times greater.

The estuary of the Severn, lying more particularly in the course of this storm, the parts bordering on that river suffered much from the breaking in of the sea. The country for a great extent was inundated, the vessels driven upon the pasture land, and many thousands of sheep and cattle drowned.

The spray of the sea carried far inland in such quantities as to form little concretions or knobs of salt on the hedges; and at twenty-five miles from the sea, in Kent, it made the pasture so salt that the cattle for some time would not eat it. The total damage done by this great storm was considered to have exceeded that of the great fire of London!

Such was one of the greatest, if not the most violent visitations of "the great November atmospheric wave" to the metropolis and the country at large—ushered in, as we are told, by a protracted wet period, and the prognostic—if such it be—may just as well be remembered in our wet years, as in the present, whilst we write.

We have drawn attention to the great fall of the barometer during the height of these equatorial tempests, suggesting its cause. In the St. Kilda cyclone of October, 1860, in the "Camilla" typhoon of the same

date nearly, the " Royal Charter" gale of October,
1859, that of the November following, and in others,
the barometer fell at the rate of nearly a *tenth of an inch
an hour before* the shift of wind occurred; previous
to which it ceased falling, then began to rise, and while
the violence of the tempest prevailed from the northward, *rose as rapidly* as it had previously fallen. Another
memorable instance of these alternations of fall and rise
occurred on the 9th of February, 1861, on the Irish
coast. That storm filled many with amazement; for
ordinary readers of the barometer, seafaring persons
especially, regard the rising of the mercury as necessarily indicating fair weather, and its falling as indicating
storm and rain,—whereas, in many localities visited by
that terrible storm, the barometer for three days before
it burst forth indicated to the minds of those who consulted it nothing but fair weather! In fact, the barometer began to rise just *before* the gale set in from the
northward, rising more rapidly as it blew stronger.
And why? Because the polar current of wind was
rushing, like a torrent of any other fluid, towards the
place of *low* barometer (lower pressure, higher temperature) where a comparative deficiency existed, which
it not only filled up, but having once acquired impetus
or momentum, filled up to a height above that which is
usual at the mean sea-level, or about 30 inches. As
Admiral FitzRoy explains it, the fall of the barometer,
with *unusually high temperature,* expressed as plainly as
by words—*southerly wind, with rain.* But that wind,
though it blew hard, did not restore the level or equilibrium of the atmosphere; and, therefore, until the
polar current approached, the barometer continued to
fall, or oscillated whilst low. Directly, however, the

northerly wind's approach was felt, the barometer began to rise (*the thermometer* had already fallen), and both thus continued to move in opposite directions—one rising for the northerly *direction* of the wind as the other fell for its *temperature.*

Finally, in the *great storm*, there is mention of " meteors and vapoury fires." The same has occurred in other tempests, especially during the months of August and November. During the great Barbadoes hurricane, August 10, 1831, fiery meteors fell perpendicularly from a vast height. This fact seems to connect these tempests with the passage of these meteors (shooting stars, etc.), at those dates, according to popular belief, and as advanced at the present day by M. Coulvier Gravier, who maintains that these fiery meteors are our weathercocks and anemometers in the upper regions of the sky.

THE MONTHLY METEOROLOGICAL MEANS OF LONDON.[1]

	Barom.	Bar. Range.	Temp.	Humid.	Rain.	Evapo.
January......	29·95 in.	1·43 in.	36·02°	91	1·56 in.	0·62 in.
February ...	29·92 ,,	1·36 ,,	39·75	89	1·45 ,,	0·84 ,,
March	29·93 ,,	1·20 ,,	42·96	83	1·36 ,,	1·64 ,,
April	29·91 ,,	1·09 ,,	47·57	77	1·55 ,,	2·52 ,,
May	29·95 ,,	0·95 ,,	55·26	75	1·67 ,,	3·99 ,,
June[2].........	29·97 ,,	0·84 ,,	60·68	76	1·98 ,,	3·75 ,,
July	29·95 ,,	0·80 ,,	63·17	77	2·44 ,,	4·24 ,,
August	29·97 ,,	0·94 ,,	62·78	78	2·37 ,,	4·03 ,,
September...	29·88 ,,	1·06 ,,	57·00	80	2·97 ,,	3·03 ,,
October	29·94 ,,	1·27 ,,	50·37	83	2·46 ,,	2·13 ,,
November...	29·83 ,,	1·37 ,,	50·44	87	2·58 ,,	1·17 ,,
December ...	29·95 ,,	1·28 ,,	40·09	92	1·65 ,,	0·62 ,,

[1] Drawn up from Daniell's ' Climate of London.'

[2] During this month the energy of the sun's beams continues to increase, though its maximum effect is a little less. The temperature of the air does not attain its maximum till the two following months.

This arrangement, as Daniell observes, must have an extremely important influence upon the fructification of the vegetable kingdom; and the horticulturist and botanist would do well to attend more particularly than has hitherto been done, to the different modifications of heat of radiation and heat of temperature. Experience has suggested many practical precautions and artifices evidently connected with this subject, and it is almost certain that a scientific attention to these particulars would tend much to the benefit of the art of gardening. "The force of radiation from the earth, I have once observed in this month to be 17°, the greatest effect that has ever come under my notice; its mean amount is 4·95°. As connected with the subject, it is worth while to notice that there is not a single month in the year in which vegetation, in particular situations, is not exposed to a temperature below the freezing-point. The two hottest months are July and August, and even in them the radiant thermometer descends to 31·10° and 29°. Thus, a plant might be so situated, in the month of July, as to undergo all the changes of heat from 130° to 31·10°!" (DANIELL, *ubi suprà.*)

CHAPTER XVI.

HOW TO INTERPRET THE BAROMETER AND THE THERMOMETER.[1]

THE lower animals get an idea of coming weather from their sensations; but for the most part man relies upon the instruments which his intellect has devised, and which are, so to speak, substitutes for the instinct of the former. No single weather-instrument is absolutely sufficient in itself, even for ordinary indications, unless it be the hygrometer; but even with this instrument we are obliged to interpret signs by the help of laws established by means of the barometer.

The barometer tells us the *pressure* or *weight* of the air at any given time. Now, what does this mean, and what has it to do with coming weather? The chief difficulty of this question is explained by what happens in a common pump. Water can be raised, or made to rise, in a common pump from 32 to about 35 feet in height, being kept there by the pressure or weight of the air on the water in the well. Thus the weight of a column of air of equal diameter is balanced by the column of

---

[1] A portion of this chapter appeared in the 'Mark Lane Express.'

water in the pump; which is thus a barometer, however imperfect; and if there were no spout to it, and the upper part were of glass and graduated in feet and inches, we should see that whenever the pump is kept in action by moving the handle, the water would rise to different heights in different weather, precisely like the mercurial barometer, remaining at a certain height, according to the weather, and unable to get higher, in spite of continued pumping. The imperfection of the pump as a barometer, consists in the fact that the air above the column of water cannot be prevented from exerting *some* pressure, the space not being a perfect *vacuum*, or exhausted of all air.

Now it was known that mercury or quicksilver is fourteen times heavier than water, and therefore it was supposed that $\frac{1}{14}$ part of 35 feet of mercury would in like manner be an equipoise or balance to the column. Accordingly a certain philosopher took a tube of glass, closed at one end, filled it with mercury, and then turning the open end into a small cistern or cup containing mercury, he found that part of the fluid, when held perpendicular, flowed out, but that the other part remained standing at a certain height, just like the water in a pump; this height being noted on the glass tube, and various alterations in the height appearing from time to time, at length, after a long series of observations, it was found that the pressure of the atmosphere, when least in foul and stormy weather, is much about equal to the weight of twenty eight inches of mercury, and in fine and serene weather that it will support about thirty-one inches of the fluid; between which extremes all the changes that happen in the weight of the air near the surface of the earth are generally found to exist. Such

is the nature of the barometer, and such were the experiments that led to its invention. It will now be evident
what is meant by the pressure or weight of the air and
its connection with the weather and its changes. When
the air is dry, deficient in vapour, or calm, it is *heavier*
than usual, and presses more upon the mercury in the
well or cistern of the barometer (concealed by the woodwork), and forces up the mercury ; then the mercury is
said to stand *high*. When the air is moist or filled with
vapour or windy, it is lighter than usual, and the pressure consequently diminishing on the mercury in the
cistern, the column in the tube falls, and is said to be
low. Now all these changes are observed in different
kinds of weather ; and as the barometer shows them, for
the most part, *before* the changes actually come on, we
see how it can be used as a weather-instrument.

In a *thermometer* the mercury is sealed up from the
air entirely, instead of being open to it at the bottom
in a cistern, as in the barometer, and the mercury rises
or falls as the varying temperature or degree of heat expands or contracts it. The heat of the air passes through
the glass tube to the mercury and causes it to expand,
whilst by the effect of cold it contracts ; and so the
thread of mercury rises or falls with heat or cold respectively.

With respect to the kind of barometers in which farmers should confide, it cannot be too forcibly inculcated
that the ordinary wheel barometer is objectionable. It
easily gets out of order, and rarely gives accurate indications of the *rate* of rise and fall, which are essential as
warnings against bad weather. Either the aneroid or
the pediment mercurial barometer should be used by
farmers, in order to prevent surprises by bad weather.

If the mercury fall ever so little between eight in the morning and noon, it indicates decreasing pressure and consequent bad weather, far more surely than even a greater fall would do if occurring between noon and four o'clock in the afternoon. Hence the importance of consulting the barometer at those early hours for the day's probable weather ; because, according to experience, the barometer naturally rises between 8 A.M. and noon, and naturally falls (both in a slight degree) between noon and 4 P.M. So if it falls when it should naturally rise, there must be some great cause for it.

But there is another cause which will make the barometer fall, without, however, the least chance of being followed by bad weather. This is owing to air getting into the vacant space above the column of mercury. All may go on tolerably well whilst the temperature of the room remains moderate ; but in the evening, when a brisk fire perhaps has been blazing, the air above the column expands and drives down the mercury. If the farmer then consults his barometer before giving his orders for the next day, it is evident that he will be far out in his reckoning.

We believe that many barometers are in this condition, and they should be occasionally tested for it, by canting them at an angle sufficient to drive the mercury up to the top of the tube. If then you do not hear the metallic clink of the mercury, there is air in the space, and you should go on canting the instrument in like manner until you hear the clink, after thus driving out the air.

As the higher we ascend from the sea-level the less the pressure of air above us, it is evident that the column of mercury will get shorter and shorter the higher it is fixed above the sea-level. It differs or stands lower about *one-tenth* of an inch for each hundred feet of height directly upwards above the sea, its average height being 29·95 inches at the mean sea-level in England. Allowances must therefore be made for barometers on high land or in buildings, each elevation having its own line of pressure or " mean."

Indications of approaching change of weather, and the direction and force of winds, are shown less by the *height* of the barometer than by its falling or rising. A *rapid* and considerable fall is a sign of stormy weather, with rain or snow. The wind will be from the northward if the thermometer is low *for the season ;* from the southward if the thermometer be high.

The average temperature of the seasons for the locality should therefore be known and remembered in consulting the barometer—on such occasions especially ; indeed, it is never safe to draw conclusions from the barometer alone, without considering the temperature and the state of the air as to moisture, as shown by the hygrometer. Many errors in prognosticating result from the neglect of this precaution.

The average temperature of the seasons for every locality can be easily ascertained upon inquiry. The following is the average at the three following places :—

	Spring.	*Summer.*	*Autumn.*	*Winter.*
Greenwich	48·4	60·3	49·4	37·8
Liverpool	46·2	57·6	49·1	40·5
Edinburgh	45·0	57·1	47·9	38·4

The chief maxims relating to the weather in connection with the barometer, with which the farmer and the public in general are specially concerned, are as follows :—

1. If the barometer has been about its ordinary height, say 30 inches at the sea-level, and is steady or rising, whilst the thermometer falls and dampness becomes less, then north-westerly, northerly, or north-easterly wind, or less wind, less rain or snow may be expected.

2. If a fall take place with a rising thermometer and increased dampness, wind and rain may be expected from the south-eastward, southward, south-westward.

3. In winter, a fall with a low ~~barometer~~ *thermometer* foretells snow.

4. A fall of the barometer, with unusually high temperature for the season, will be followed by a southerly wind, with rain; and during the gale the barometer may begin to rise, and be followed by another from the northward; but then the thermometer will fall for *change of temperature,* and show the *direction* of the coming wind.

5. But northerly winds will follow a fall of the barometer at all times if the thermometer be low (for the season), and southerly wind, if the thermometer be high for the season.

6. The barometer is *lowest of all* during a thaw following *a long frost,* because the air, which had been much dried by the frost, absorbs the moisture of the fresh warm current of wind from the south or south-west; and secondly, the air, having been much condensed by frost, is suddenly expanded by the warm wind which is introduced.

7. In like manner the glass falls very low with south and west winds in general, because these winds are generally laden with vapour, and vaporized air is lighter than dry, in which there is a deficiency of air and therefore of pressure. Such falls with a southerly wind in the autumn are very often attended by thunderstorms.

8. If the barometer falls with the wind in the north we must prepare for weather of the worst description, —rain and storms in summer, snow and severe frosts in winter and early spring.

9. A rapid rise in winter, after bad weather, is usually followed by clear skies and hard white frosts. The pressure of the air is increased by cold.

10. During frosty weather, if the barometer falls it denotes a thaw; but if the wind goes again to the north, the mercury will rise, and the frost set in again.

11. During broken and cold weather in the winter with northerly winds, a sudden *rise* of the barometer foreshows a change of the wind to the southward, with rain.

12. If, during a northerly and easterly wind in winter, the barometer rises slowly, expect snow or cloudy weather.

13. In a continuous frost, if the mercury rise, it will certainly snow.

14. Whilst the barometer stands above 30°, the air must be very *dry* or very *cold*, or perhaps both, and no rain may be expected. For, if it be very *dry* it will *absorb* moisture, and not part with it in rain. If the air be very cold, it is so much condensed that it has already parted with as much moisture as it can spare.

15. Of course a great rise in summer means dry and warm weather, and if this be of long duration the question is, how will it end? If a sudden fall, of two or three tenths occur, we must prepare for a storm of rain, or thunder and rain. This usually follows a period of unusual heat, unless northerly and easterly winds are to succeed, with drought.

16. Thunderstorms are not always foretold by the

barometer; indeed, the barometer cannot indicate electricity, as some of the barometer-maxims seem to infer.
We must consult the clouds and our feelings for thunder.
The barometer falls, but not always, on the approach of
thunder and lightning. It is in very hot weather that
the fall of the mercury indicates thunder. Thunder
clouds rising from north-eastward *against the wind,* do
not usually cause a fall in the barometer, simply because they are borne onward by a *polar current,* which
is dense. An approaching thunderstorm is indicated
by a rapid decrease of the daily *evaporation* during hot
weather.

17. A rising barometer—with a southerly wind—is
generally followed by fine weather; but then it will be
generally observed that a change of wind occurs at the
same time, or very shortly after.

18. During stormy weather the mercury will be seen
to rise and fall continually, and to be in a state of general agitation. Whilst this lasts no hope of good
weather can be entertained.

19. Although some rain may occasionally fall with a
high barometer, it is usually of trifling amount, and of
short duration.

20. When the barometer *stands* very low indeed, there
will never be much rain; although, on the other hand, a
fine day will seldom occur at such times. The air must
be very warm or very moist or both, and so there will
only be short heavy showers, with sudden squalls of
wind from the west. For, if the air be very warm it
will have a tendency to imbibe *more* moisture, and not
to part with what it has. If the air be ever so *moist,*
rain will never fall till cold air has been introduced to
condense the vapour, and the moment the *cold* air is

introduced the barometer will rise, because it will condense the air.

21. A sudden fall of the barometer, with a westerly wind, is sometimes followed by a violent storm from N.W. or N.N.E.

22. In summer, after a long-continuance of fair weather, the barometer will fall gradually for two or three days before rain comes ; but if the fall be very sudden, then a thunderstorm is at hand.

23. When the barometer is *high*, dark, dense clouds will pass over the sky, without rain ; but if the glass be *low*, it will often rain without any appearance of clouds.

24. If wet weather happens soon after the fall of the barometer, there will be little of it. In fair weather, if the glass falls much, and remains low, expect much wet in a few days, and probably wind. In wet weather, if the glass continues to fall, expect much wet.

25. The barometer sinks lowest of all for wind and rain together ; next for wind, except it be an east or north-east wind.

26. Instances of fine weather with a low glass occur, however, rarely ; but they are always forerunners to a *duration* of wind or rain, if not both.

27. Our storms are generally announced by a fall of the barometer, and a tendency of the wind towards east and south ; the return of fine weather, by a rise and a pretty strong west wind, apparent in the motion of the *clouds* before it is felt below.

28. A great and sudden change, either from hot to cold or from cold to hot, is generally followed by rain within twenty-four hours ; because, in the change from hot to cold, the cold condenses the air and its vapour, which being condensed falls in rain ; and in the change

from cold to hot, the air is quickly saturated with moisture, and as soon as night comes on, the temperature is lowered again, and some of the abundant moisture falls in rain.

29. When heat rapidly follows cold, the evaporation, which was checked by cold, is carried on very rapidly, in consequence of the diminished pressure of the air by heat. The less the pressure of the air, the more rapid the evaporation of moisture.

30. The barometer varies most in winter, because the *difference* of temperature between the torrid and temperate zones is much greater, and produces a greater disturbance in the state of the air. It varies least in summer, because the temperature of our island is so nearly equal to that of the torrid zone, or hot regions of the earth, that its state is not much disturbed by interchange of currents.

31. Heat and cold do not of themselves affect the barometer, but because cold weather is generally either dry or rough, with north-east winds, the air being denser and heavier—therefore the mercury *rises* in cold weather; but in warm weather the air is often moist and less dense, with south-west winds, and therefore lighter, and so its pressure is less, thus causing the mercury to fall.

32. If the top of the column of mercury be convex, or higher in the middle than at the sides, it is rising; if lower, or concave, it is falling. This is caused by the attraction of the glass-tube in contact with the mercury.

Such are the chief rules and maxims of the barometer with reference to agricultural pursuits, the seasons, and the crops. No attention should be paid to the words engraved on certain barometers, such as "set fair,"

" rain," etc. These expressions have done much injury, since many persons have been misled by them ; and they should be discontinued by general consent of the instrument-makers, together with the total abolition of the banjo or wheel barometer.

Moreover, the words " fair," " changeable," etc., should be removed and placed more fitly, and with some reference to the elevation above the sea at which the barometer is used. It is clear that *at the sea level*, the height of 29·80 will more truly represent the *medium* or limit of fair and wet weather, than 29·50, at present marked "changeable." In employing the instrument as a weather-glass it is needful to the success of the observer to attend to it daily, and to watch the approach of the quicksilver to the *true limit of the fair and wet indications*, and its reading from this in ascent or descent ; with attention to past as well as present circumstances. Fair weather and sunshine may often be found to accompany a low state of the barometer ; but they are not in that case to be depended on. Continued movements in either direction may be safely taken as indicative of a corresponding duration in the weather they imply,—as in the weather following the change at the close of July, 1866. When the mercury in its downward course has passed the true changeable point above given, *rain*, if it has not fallen, is to be expected *in its rising again above it ;* after which, the fair weather indicated by that rise is found to prevail.

The most difficult weather of all for the farmer, when his labour requires a certain degree and continuance of dryness, is that in which the quicksilver makes frequent short movements, in ascent and descent, still keeping about the point which answers to the limit above mentioned ; and *which point it is therefore necessary to settle accurately for him,* according to the elevation at which he is placed above the sea.

One word respecting the "storm glasses" may be required. They consist merely of a solution of camphor, saltpetre, and sal-ammoniac, in diluted spirits of wine. These toys are useless in prognosticating the weather, being only affected by light or temperature, and only when the mixture is accurately made—as it seldom is : hence the very great difference observed in the different "storm glasses" of the shops. As the effect of the chemical or actinic rays of light and temperature differs with the antecedents of any weather, no doubt these glasses may be sometimes "correct" in their indications : hence the belief in them, if it really exists. We need not state that they cannot indicate the *pressure of the air,* upon which depend all states and changes of the weather; and this must be discovered either directly by the barometer, or indirectly by the hygrometer.

The atmospheric pressure varies with the direction of the wind; the barometer is everywhere very high when the wind blows between the east and the north, and often very low when it comes from a point between the south and the west; its height generally varies very regularly between these two extremes, except during the winter months, when a rise or fall of several tenths corresponds in effect to that from a single tenth in spring and summer.

In certain places, however, anomalies are found; thus at Vienna and Buda, the pressure is very feeble with *east* winds; and at St. Petersburg, the *minimum* almost coincides with the N.W.

These anomalies have not yet been well explained, for

they are not derived merely from the continental positions of these towns, since the results obtained at Stock-
holm and Moscow agree with the laws which regulate
western Europe. The only difference consists in the
fact that the oscillations are a little narrower in the interior of the continent than on the west coast.

Analogous laws are found in other countries,—only
the wind that corresponds to the *maximum* barometric
height varies according to the position of these points in
relation to *Europe*. Thus, in the United States, the
barometer is highest with N.W. winds, and lowest with
S.E. ; it is the same at Pekin, in China.[1] On collecting
various facts together, we conclude that *the barometer
attains its maximum when the winds blow from the north
and from the interior of continents,—the minimum when
they come from the equator or the sea in our latitude.*

Hence, the daily announcements in the *Weather
Table* of the Meteorologic Office, respecting high or low
barometric pressure in this or that region, must not always be interpreted—as seems to be done—with reference to the rules which guide the rise or fall of the
mercury at home. This fallacy may have been the
cause of numerous failures in the forecasts. The further
we go for signs from the barometer, the greater should
be our care to interpret them according to the rules
established for the instrument in each region. The
axiom that the wind will blow " from high to low barometer" is absolutely true ; but the *direction* of the wind
must be different in different regions, if different winds
regulate the rise and fall in different regions. According to Dr. Buys Biot, the eminent Dutch meteorologist,
from observations in Holland, it appears that a high

[1] Kaemst, *ubi suprà.*

reading of the barometer at northern stations is followed by an easterly wind, and that a high reading of the barometer at southern stations is followed by a *westerly* wind.[1]   So much for the " *areas of pressure.*"

The intelligent observation of the thermometer should always accompany that of the barometer. We must never forget that temperature is the starting-point of all meteorological phenomena, since, in a manner, it includes even the causes termed electrical. The following rules are quoted by Orlando Whistlecraft, a meteorological veteran of considerable popular celebrity.

1. In winter, whenever the thermometer stands about 34° by night (which is neither quite freezing nor yet mild), and when by day about 45° (which also is a medium between cold and mild), *it is a sign of much rain.*

2. In summer, when the thermometer rises gradually with the barometer, it is more certain that it will be fine weather than if the barometer only be rising.

3. If in summer the thermometer stands *low*, with S. or S.W. wind, be sure of much wet; but if *high* with those winds, it will soon be clear sky and great heat.

4. If the thermometer be unusually high during the night in summer, there must be thunderstorms rising somewhere in England at the time.

5. If the thermometer be high during winter nights, be sure of high winds in a few hours.[2]

The use of a Terrestrial Radiation Thermometer, for indicating the lowest temperature of terrestrial radiation

---

[1] Quarterly Report of Meteorol. Soc. of Scotland, June, 1863.

[2] Whistlecraft's 'Weather Almanac' for 1867, decidedly the most interesting of the kind.

or the *minimum temperature of vegetation* cannot be too forcibly recommended to agriculturists and horticulturists, especially the latter. By its continued indications not only may the results of the season be inferred beforehand, but they may suggest the adoption of appropriate means to counteract untoward seasons. A series of such annual records would be a most instructive meteorological study, calculated to be ultimately most advantageous to farmers and horticulturists.

If Casella's admirable Mercurial Minimum Thermometer be not too expensive for the purpose, it seems best adapted for a series of such observations in the hands of a diligent and intelligent observer.[1]

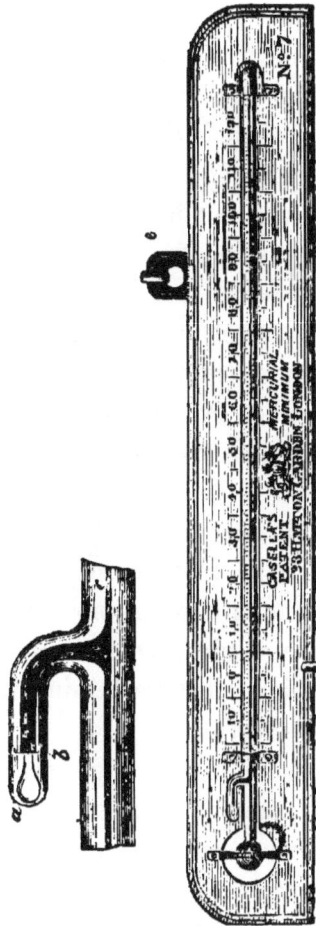

The idea of this instrument is a beautiful scientific

[1] The general form is shown in the figure ; *d* being a tube with large bore, at the end of which a flat glass diaphragm is formed by the abrupt junction of the small chamber (*a b*), the inlet to which at *b* is larger than the bore of the indicating tube. The result of this is that, having set the thermometer, the contracting force of the mercury in cooling withdraws the fluid in the indicating stem only, whilst on its expanding with heat, the long column does not move, the increased bulk of mercury finding an easier passage through the larger bore into the small pear-shaped chamber attached.

application of the adhesive property of mercury for glass *in vacuo;* and the fact that where two tubes are united to one bulb, this fluid will rise, by expansion, in the larger, and recede, by contraction, in the smaller tube. The result is the invention of probably the first practical mercurial instrument known to register past indications of cold without having or forming any separate index, or in which the adhesive property of mercury for glass as a fixed point has ever been employed. It is a wonderful instance of how small and delicate a force suffices to produce results the most important; and the invention is highly creditable to Mr. L. M. Casella, the son of the well-known scientific instrument-maker of Hatton Garden.

Not only for agricultural and horticultural purposes does this instrument seem admirably adapted, but also for the investigations of meteorology more especially, wherein great sensitiveness and perfect accuracy on all occasions in ascertaining the *daily range* are absolutely necessary.

Owing to the high price of this instrument, Mr. Casella has introduced to the notice of the Meterological Society a valuable modification of the ordinary standard minimum thermometer, without increasing the cost. In this, as in all other good arrangements, the plan is very simple—the bore of the stem being merely smaller than usual and the bulb small in proportion, but extended to a much greater length and bent up so as to concentrate the exposed surface as much together as possible. The result obtained is the indication of changes of temperature in less than half the time required in any other arrangements that have come under our notice.

CHAPTER XVII.

SOME years ago a collection of the popular weather prognostics of Scotland was made by Dr. A. Mitchell, and that public-spirited nobleman the Marquis of Tweeddale offered a prize of a gold medal, a piece of plate, or twenty guineas, for the best scientific explanation of them.[1] The science of meteorology is much indebted to the distinguished and liberal patronage of the Marquis, but this idea of directing scientific attention to our popular weather-axioms deserves special commendation. The popular weather-prognostics or signs form one of the most curious and interesting branches of meteorology. They are very similar among all nations, and have been so in all times. We find them recorded or alluded to in the earliest writers, sacred and profane, —in the Bible and in the ancient poems of Greece and Rome. Upwards of two thousand years ago the Cilician poet Aratus devoted his muse to the subject, and his popularity may be inferred from the fact that he was the poet so pointedly alluded to by St. Paul, saying to the

---

[1] We are informed that the prize was duly awarded, but it does not appear that the successful explanation was published.

Athenians :—" As certain also of your own poets have said, For we are also His offspring."[1]

The Roman poet Virgil abounds with such weather-prognostics, but he merely copied Aratus on this subject, just as he copied Homer in his other compositions.

Aratus discourses largely on the signs of weather from the appearance of the stars, the sun, and the moon, just as they are now repeated in popular tradition—even mentioning " the moon lying on her back," as a prognostic of a south wind. He speaks of the shooting stars or meteors, which, long before M. Coulvier Gravier, the French meteorologist, Aratus supposed to indicate the direction of coming wind.[2]

Among the signs of rain he instances as we do, " the pond and sea birds insatiately dipping into the waves, eager for moisture; the swallows fluttering around the banks, untiringly striking the stream with their breasts."

The signs from the " querulous frogs," from lowing and sniffing oxen, biting flies, etc., just as we quote them, are also recorded in this very ancient poem ; even the crowing of cocks at unusual hours, the crowding of the poultry near the homestead, ants carrying their eggs to shelter, the *rapid growth of mushrooms overnight*— all signs of *abundant moisture in the air,* and therefore of rain.

[1] Acts xvii. Τοῦ γὰρ καὶ γένος ἐσμὲν are the words of Aratus. *In Phainom.* The fact of his being thus quoted not only shows that he was a popular poet but also the tact of the Apostle in alluding to him as such. Aratus flourished about the year 278 B.C. There are two of his poems extant—the " Phænomena" and the " Diosemeia" or Prognostics. Aratus copied Hesiod, Aristotle, and Theophrastus on the subject, but he has evidently given much from his own observation. The *Diosemeia* is a beautiful poetic treatise on the signs of coming weather.

[2] " Δειδέχθαι κείνοις αὐτὴν ὁδὸν ἐρχομένοιο Πνεύματος." *Diosem.* 194.

In these allusions by the ancient poets, as Dr. Mitchell observes,[1] there is nothing of the supernatural or mysterious. They speak of them as simple deductions from the experience or observation of those whose occupations led to the looking for signs of coming storms. It is true they may now be called traditional, but as every generation has had the opportunity of testing their accuracy and value, unless they had been found to contain some measure of truth they would soon have ceased to be handed down. It must not be forgotten that they regulate the affairs of every day life, and lead to loss or gain—a test which soon deprives traditional error of its vitality; and this is only a little less slowly its fate when it is linked or mixed up with religious belief.

Dr. Mitchell's collection of Scotch weather-prognostics has its complete correspondence in those of every other nation; we shall adopt his classification, and begin our explanations with—

I. PROGNOSTICS FROM HILLS OR MOUNTAINS.

In Kilconquhar and Elie, a cloud resting on Largo Law is a sign of coming rain, while one on Kellie Law presages clear weather.

> " When Largo Law puts on his hat,
> Let Kellie Law beware of that.
> When Kellie Law gets on his cap,
> Old Largo Law may laugh at that."

Take another—rather more poetically expressed :—

> " There's a high wooded hill above Lochnau Castle,—
> Take care when Lady Craighill puts on her mantle.
> The Lady looks high and knows what is coming—
> Delay not one moment to get under covering."

In Dumfries, Kirkpatrick Fleming, and several of the

[1] ' Popular Weather Prognostics of Scotland.'

T

intervening parishes, the rolling of clouds landward, and their gathering about the summit of Criffel, is regarded as a sure indication of approaching foul weather.

Heavy clouds on Skiddaw, especially with a south wind, are a sign of coming rain to the farmer of Kirkpatrick Fleming.

The natives of Gigha and Cara anticipate rain when the Paps of Jura (to the north-west) are capped, and if the cloud be white, they expect wind with it.

When a thick cloud on the top of a hill is observed to be in motion, the certainty of the prognostic is regarded as increased.

Such are some of the hill-prognostics quoted by Dr. Mitchell, and the same explanation will answer for all.

The presence of clouds on the summit of mountains is noted in all countrise, especially in the Pyrenees and the Alps, as a sign of coming rain; and precisely the same form of expression is used in the saying; such-and-such a mountain "has put on his hat," is everywhere used for "we shall soon have rain."

If Dr. Mitchell had verified any of these prognostics and defined the *species* of cloud in every instance, together with its alleged sign—whether fair or foul—each would have given its own explanation. A cloud on the summit of a mountain or hill or other high place will bring rain, provided it is a *cirrus*, for *cirri* are the first to be surrounded by *cumuli* when these begin to form, and it is the union of these clouds that produces rain.

So much for the sign of rain from clouds on the hill-tops; but how can they also indicate the contrary, as in some of these sayings? Accurately enough—if it be a different species of cloud—a *cumulus*. When the sky is clear above the elevated summits, as will be evident from its beautiful blue tint, *cumuli* will loiter there sometimes for several days, without rain, and may finally disappear under the influence of the sun's rays, or be wafted away by the wind during the night. So it is the cirrus on the hilltop that brings rain, and the cumulus that secures fine weather. On Skiddaw, for instance, it must be a cirrus, to which the *south wind*, as stated, must bring its special cloud—the cumulus; and if the cirrus be on Kellie Law to the *east*, of course there will be little chance of rain at Largo Law in the *west*, as stated in the saying.

II. PROGNOSTICS FROM MISTS AND FOGS.

In the evenings of autumn and spring, vapour rising from a river is regarded as a sure proof of coming *frost*.

Hazy weather is thought to prognosticate frost in *winter*, snow in *spring*, fair weather in *summer*, and rain in *autumn*.

Thin, white, fleecy broken mist slowly ascending the sides of a mountain whose top is uncovered, predicts a fair day.

> "When the mist creeps up the hill,
> Fisher, out and try your skill;
> When the mist begins to nod,
> Fisher, then put by your rod."

That is, if the mists ascend—fair weather; if they descend—foul.

White mist in winter indicates frost.

Such are the popular sayings regarding fogs and mist; and a little explanation will show that they are founded on fact.

The formation of mist, which is aqueous vapour, shows that an additional quantity of vesicular vapour is being added to that already contained in the atmosphere, and therefore that the volume of the clouds will be augmented. Thus do mists secure the primary conditions of rain in autumn, at which season the conditions of cloud (cirrus and cumulus) and temperature readily effect the condensation into rain.

Mist in winter shows that the air is getting colder than the surface of land or water, and therefore frost may be expected.

The same condition in spring, when rain or moisture predominates, will naturally convert rain into snow.

If there be mist in summer, especially when the wind blows from the north or east, the heat of the sun dissolves this vapour and prevents clouds from forming; hence there will be no rain, but fair weather.

If the mist ascends a mountain whose top is uncovered, that is, not wooded, there can be no rain, for it is round the tops of *wooded* mountains that the *cirrus* loves to hover in wait for the vesicular vapour in any shape ascending from the plains and valleys. It is, as before stated, the union of the icy cirrus and the warmer moist cumulus that produces rain.

A *white* mist is caused by there being no clouds to blacken it with their shadow, and if this occurs in winter it shows the first condition of frost—a *clear sky*. Hence, the probability of the saying. We need not state that a cloudy sky makes a *black* mist, and clouds in the sky reflect back what heat may be radiating from the earth, and so tend to prevent frost.

On the other hand, a black mist in summer portends rain, because it is accompanied by dense clouds above; whilst a white mist in summer indicates fair weather, because there are no clouds to give rain.

When mists rise and envelope the hills, the atmosphere must get highly charged with aqueous vapour. The prognostication does not extend to the calm days of November and December, when general fogs occur in consequence of the temperature of waters and damp situations, being higher than that of the ground and the air above it. Great fogs have generally a high barometer and much colder atmosphere. Fogs will occur after severe frosts, at the commencement of a thaw, when a warm south-west wind has given out part of its moisture, owing to the surface having been previously rendered very cold.

III. APPEARANCES OF THE SKY.

1. A small cloudless space in the north-east horizon, especially if the clouds generally are moving to the south or south-west, and if the weather has previously been wet, is regarded all over Scotland, among seamen and landsmen, as a very certain precursor of fine weather, or a *clearing up.*

This is what the sailors call just enough of *blue sky* to wipe one's face with. It denotes that the upper regions are clear, and of course if the clouds are moving towards the south or south-west—the points whence we generally have wet—"a clearing up" may be expected.

2. In winter, when the sky about midday has a greenish appearance to the east or north-east, snow and frost may be expected.

We infer that the wind must be in the east or northeast on such occasions, and then the *greenish* vapours of the air show that they are already *condensed into clouds,* which will soon give rain in summer, or snow in winter. Condensed vapours refract yellow or greenish rays, because the beams of light meet with more resistance from them, so that those which are bent down to the eye are the most refracted, yellowish or greenish.

> 3. "The evening red, and morning grey
> Is the sign of a bright and cheery day;
> The evening grey and the morning red,
> Put on your hat, or you'll wet your head."

Such is the Scottish; in England we say—

> "Evening red and morning grey
> Set the traveller on his way;
> Evening grey and morning red
> Bring down rain upon his head."

This is a very old adage; indeed, it is alluded to by our Lord (Matt. xvi. 2, 3), "When it is evening ye say, It will be fair weather, for the sky is red;" and again, "In the morning ye say, It will be foul weather to-day, for the sky is red and lowering."

The inference from this would be the implication of a contradiction; but this is not the case; the consequence is perfectly deduced. A *red* sunset shows that the vapours are not condensed into clouds, but only on the point of being condensed, in which state, we know, they bend the *red* rays of the sun towards the horizon, where they tint the floating clouds. If not condensed, there can be no rain.

The *red* sun-*rise*, on the contrary, indicates a wet day, because the higher regions of the air are laden with vapour on the very point of condensation, as before said, which te rising sun cannot disperse. So, there will be rain, if, as stated in the Scottish axiom, "it holds out till the sun is fairly above the horizon," and and certainly if it "last for some time."

4. When, in the morning, the dew is heavy and re-

mains long on the grass, when the fog in the valleys is slowly dissipated and lingers on the hillsides, when the clouds seem to be taking a higher place, and when loose cirro-strati float gently along,—serene weather may confidently be expected for the greater part of that day.

On Dr. Mitchell's authority, we of course accept this as a "popular" weather prognostic; but we may be excused for saying that it is rather long-winded and too "categorical" for a *popular* adage. We do not believe that "cirro-strati" have as yet got among the "populars," and so we had better take the axiom as a compound, and consider each of its elements. The concluding words, "greater part of that *day*" simplifies the question. Nights with abundant dews may be considered as foretelling *rains*, because they prove that the air contains a great quantity of aqueous vapour, and that it is near the point of saturation; but if, as stated, a large quantity of dew has fallen, there is, so far, a diminution in the quantity of vapour in the air, and therefore "serene weather may be expected for the greater part of *that day*."

The absence of dew in the morning is certainly a sign of fine weather, for it indicates a reaction in the reflecting power of the clouds; and if there be but *one species* of clouds in the skies—cumulus or cirrus—it never rains if there be no dew in the morning.

The other elements of the prognostic—"fog slowly dissipated, etc., clouds seeming to be taking a higher place, and loose cirro-strati floating gently along," may be a consequence of the heavy dew-fall of the previous night ridding the atmosphere of superabundant moisture. But, altogether, this can scarcely be considered a "popular" saying.

5. Continuous cirro-strati gathering in unbroken gloom and also the cloud called "goat's hair" or the "grey mare's tail" presage wind.

The cirro-stratus is a combination of the icy cloud with the vesicular cloud, and therefore indicates rain, and wind from the *refrigeration* which the cold rain produces. It is the "mackerel sky" of England; and the other cloud mentioned is the *cirrus;* hence our English saying:—

> "Mackerel's scales and mare's tails
> Make lofty ships carry low sails."

6. Light fleecy clouds in rapid motion below compact dark cirro-strati foretell rain near at hand.

Again, the wording of this axiom is scarcely "popular," and if it means anything it is simply the fact that cumulus clouds in motion under the cirrus formation will produce rain,—which has been already stated.

7. When after a shower, the cirro-strati open up at the zenith, leaving broken or ragged edges pointing upwards, and settle down gloomily and compactly on the horizon, wind will follow, and last for some time.

We do not hesitate to say that no such appearance ever occurred; the entire description is contrary to the fact as connected with any cloud of the cirrus species. When the "streamers" ("broken edges") of cirrus clouds point upwards, the clouds are descending towards the cumulus, and rain is at hand; but when their streamers point *downwards*, they are ascending, and we may expect drought, with easterly winds, the usual concomitant, owing to their dryness and density.

8. When, after a clear frost, long streaks of cirrus are seen with their ends bending towards each other as they

recede from the zenith, and when they point to the north-east, a thaw and south-west wind may be expected.

It is the effect of the south-westerly current in the upper regions that points the cirrus to the north-east, as they recede from the zenith with their ends bending towards each other.

9. Cumulus clouds, high up, are said to show that south and south-west winds are near at hand ; and stratified clouds, low down, that east or north winds will prevail.

Cumulus clouds are the special clouds of southerly and south-westerly winds, and therefore if they are " high up," or prevail, their wind may be expected ; what is meant by " stratified clouds " we know not ; but a clear upper sky with an accumulation of clouds low down, would indicate greater density in the atmosphere, which is the state during east and north winds.

10. Cirrus at right angles to the wind is regarded as a sign of rain.

If the icy cloud thus lie in the path of the wind bringing the clouds of vesicular vapour, the latter cannot escape them, and so of course there will be rain by the contact or encounter.

11. The farmers of Berwickshire say that a long stripe of cloud, sometimes called by them a salmon, sometimes called Noah's ark, when it stretches through the atmosphere in an east and west direction, is a sign of stormy weather, but when it stretches in a north and south direction, is a sign of dry weather.

This must be an extensive cirrus, thus arranged from west to east by the equatorial current in the upper regions of the atmosphere, and therefore the storms of this current may be expected. If it stretches from north

to south, it shows the predominance of the polar current,
with its usual north-easter and drought.

12. Along the north shore of the Solway, from Dumfries to Gretna, a lurid appearance in the eastern or
south-eastern horizon, called from its direction "a Car-
lisle sky," is thought a sure sign of coming rain. They
describe it as lurid and yet *yellowish*, and the common
saying is—

> "The Carlo sky
> Keeps not the head dry."

The *yellowish* sky shows that the *south-castern* atmosphere is full of vapour already condensed into clouds, and
therefore rain may be soon expected.

13. In Kincardine of Monteith, and all that district
of country, the reflection from the clouds of the furnaces
of the Devon and the Carron (to the east) foretells rain
for next day.

Rainy weather is indicated by the greater distinctness
of distant objects, and their apparently increased elevation, called *looming*, which is more conspicuous over
water, and depends upon the refractive power of the air
being increased by the *aqueous vapour* in it.

When there is a deficiency of aqueous vapour in the
atmosphere, during the time that airs of different temperatures are mixing, there will be *haziness*, and objects
will not be clearly seen, and the stars will appear to
twinkle from the unequal refractions of them, before
the airs are in complete union. This is frequently per-
ceivable, especially in the spring months, and prognosticates *dry* weather.

If, however, the aqueous vapour be in sufficient quantity, the caloric transmitted by it tends rapidly to equa-
lize the temperature of the contiguous dry airs, and to

cause their speedy union; now, the distinctness of distant objects is in proportion to the quantity of it, and, consequently, to that of the probability of approaching rain.

It is the *moist air* which enables the clouds in Kincardine to transmit the reflection of the furnaces in question.

On this principle depend the various prognostications of the weather from the different appearances of the celestial bodies, particularly the moon, and also the blueness of the sky, which is often *very remarkable between clouds before rain,* at which time the sun shines very hot, from his rays not being obstructed by the mixing of airs of different densities.

In forming opinions respecting atmospheric changes, either to rain or drought, from the sun, moon, or stars, clouds, and sky, the difference between the transparent air and that of the mixing portions of it, when of different temperatures, must be distinguished from that mistiness arising from the condensation of vapour into aqueous globules; these occasion that *indistinctness* which is generally the precursor of rain.

14. The glare of the distant Ayrshire ironworks being seen at night from Cumbrae or Rothsay, rain is expected next day. Similar prognostics are common all over Scotland.

The explanation of this is included in that of the former.

15. A *mackerel* sky denotes fair weather for that day, but predicts rain a day or two after.

It is the cirro-stratus cloud which is called "mackerel's scales." Of course until the *cumulus* makes its appearance there will be no rain; but the probability is that it will soon appear;—hence the saying.

IV.—MOON AND SUN, RAINBOW, TIDES, ANIMALS,
SMOKE, AND DURATION OF FROSTS.

1. A few days after full or new moon, changes of weather from good to bad or bad to good, are thought more probable than at other times.

This axiom opens the great question as to the influence of the moon on the weather, which is utterly denied by the astronomers. It must be admitted that the facts seem more like coincidences of the moon's phases than effects of her influence alone. The phases may therefore be only contemporaneous with certain actions and reactions between the earth and the moon, and between the *Earth-Moon* and the sun. Thus, the full and new moon are critical periods, but chiefly at the equinoxes and solstices.

2. In winter, when the moon's horns are sharp and well-defined, frost is expected.

The sharpness results from the absence of aqueous vapour in the upper regions of the atmosphere; if there be no saturation of vapour there will be an absence of clouds, and consequently a clear sky, which favours frost, as before explained.

3. When the moon has a white look, and when her outline is not very clear, rain or snow is looked for.

This state indicates much aqueous vapour in the upper regions of the air, and therefore the probability of rain.

> 4. Clear moon
> Frost soon.

This is explained as under No. 2.

5. If the new moon embraces the old moon, stormy weather is foreboded.

Great confidence is placed in this old prognostic :—

> " I saw the new moon late yestreen
>   Wi' the auld moon in her arm,
> And if we gang to sea, Master,
>   I fear we'll come to harm."
>
> *Sir Patrick Spens.*

Saturation of vapour in the upper regions of the atmosphere is the cause of this phenomenon, as stated under No. 3 ; and we have observed it twice, followed by stormy rain.

Haloes and mock-suns come under the same category —all the result of superabundant aqueous vapour in the atmosphere.

> 6. " A rainbow in the morning—
>   Sailors take warning ;
>   A rainbow at night
>   Is the sailor's delight."

In England we refer this to the *shepherd ;* and the explanation is as follows :—A morning rainbow must be always in the *west,* because the sun is in the east, and a rainbow can be formed only when the clouds, which are dropping rain, are opposite to the sun ; and it indicates bad weather coming to us, because our heavy rains generally come from the *south-west ;* and therefore clouds which show the rainbow in the west are coming up with the wind, bringing rain with them.

But a rainbow at night must be in the *east,* and it indicates bad weather *leaving* us, because the sun is in the west, *from* which quarter the rain-clouds (which reflect the rainbow) have been driven to the *east.* Therefore, the storm has passed over us and is departing.

So far we have followed Dr. Mitchell's collection *seriatim,* and most of the remainder will be found explained in the following summary of popular weather prognostics :—

Change of weather to damp or rainy is indicated by a

difference in the *sound* of the wind, which is more acute
or whistling than usual, and continually rising or falling, as in the tones of an Æolian harp.   The observation
of Elijah—" There is a sound of abundance of rain "
(1 Kings xviii. 4), shows the antiquity and universality
of the knowledge of this circumstance.   Sounds of bells
and other distant objects are heard louder and more distinctly before rain, particularly when there is not much
wind, although the air may be less dense.   The greater
audibility has been attributed by some writers to the
clouds acting like a sounding-board ; but distant noises
may be heard plainer from the air in a more homogeneous state with respect to density.   During the severe
frost of the Arctic regions, as related by Captain Parry,
this was certainly the case, when the air and the surface
were of the same temperature.   If the atmosphere contains a considerable quantity of aqueous vapour, it becomes homogeneous from the rapid equalization of temperature, as above stated, and consequently foreshows
rainy or cloudy damp weather.

In calm weather the murmuring sounds from the
rustling of the leaves of the trees of woods, from hollow
places and pits, and from the sea, as noticed in our
popular sayings, indicate a change of weather, and probably rain ; because they depend upon the *increasing
rarity of the air*, which causes that which is cold and
dense to expand, from less pressure, and by its action
upon the obstacles which may impede its progress, to
produce the noise.

When the earth appears nearly dry and free from dirt
after heavy showers in summer, this is an indication of
more rain coming, because it shows there is a great heat
of surface, and it is attended by a falling barometer.

Unusually fine warm days in spring, and also in autumn, and sometimes in winter, are commonly followed by cold rainy days, if the surface of the ground be wet, or in the vicinity of shallow waters; sometimes they are followed by boisterous windy days. The air over the surface becomes heated by the ground and consequently expands, and is succeeded by colder and denser air, owing to clouds being formed after sunset. The atmosphere being saturated with moisture, the clouds give rain, which cools the ground and increases the production of aqueous vapour during the next and succeeding days, according to circumstances.

Very cold weather is generally followed by rain or snow. In consequence of the cold, the lower column of air becomes very dense, and that above so rare or thin as to cause a determination of air from the vicinity, which, at that height, is a warmer region, and charged with more aqueous vapour than the change of temperature from the sudden cooling on its passage into the colder column, although less dense, can enable it to support.

THE DURATION OF FROSTS.

Frosts are more lasting that commence with a gentle east or north-east wind ; they increase in intensity for a considerable time, when a calm takes place, the temperature suddenly diminishes, the barometer falls very much, and a thick mist arises, followed by a change of wind and a thaw.

If frosts begin during south-west gales, the wind suddenly changes from south-west to north-west, generally about an hour after sunset, and blows very strongly in this direction, attended with severe frost and sometimes heavy snow. A frost of this kind seldom continues longer than thirty-six or forty hours. On the morning of the second day the temperature gradually increases; usually before noon the wind suddenly shifts to south-west, with a falling barometer; a rapid thaw takes place, frequently accompanied by rain.

Beams of light appearing, when the rays of the sun are reflected by dense clouds, through their intervals upon floating particles of dust or mist, in a similar manner to those seen in the rays passing by means of a small aperture into a dark room, are supposed to indicate approaching rain. The dark dense clouds and the condensing aqueous vapour show that there is a large quantity of moisture in the atmosphere, and consequently a tendency to wet weather.

SIGNS FROM THE COLOURS OF THE RAINBOW.

There are various prognostications given by the greater or less bright tints of certain colours which may be seen in rainbows. These arise from the magnitude of the drops of rain, which refract them in greater or less quantities; and also from the aqueous and mineral particles which are floating in the atmosphere and prevent the passage of larger or smaller quantities of rays of different colours from the light of the sun. Some authors say that the predominance of dark red in the rainbow shows tempestuous weather; of light red, wind; of yellow, drought or dryness; of green, rain; and of blue, that the air is clearing. The frustum or part of a rainbow only seen, is supposed to be an indication of squally showers. Aqueous vapour in different quantities and in different conditions refracts different rays of light; hence, these prognostications.

We have explained the prognostic from the rainbow in the morning or in the evening. We may observe in addition, that a rainbow in the morning may arise from the temperature of the atmosphere during the night not having been sufficiently depressed (owing to radiation being hindered by clouds) to condense the aqueous vapour; consequently it will increase during the day and produce rain. The rainbow in the evening will show that condensation, arising from decreasing temperature, has already commenced, and will most probably clear the atmosphere of a great part of its moisture in the course of the night, by which means rain will be prevented during the ensuing day.

SIGNS FROM THE MOON.

An old adage says, " Pallida luna pluit, rubicunda flat, alba serenat."

" A pale moon rains, a red moon blows, a white makes fair."

That is, if the moon appears pale, it indicates rain; if red, wind; and if clear, fair weather. The paleness of the sun or moon, particularly at rising or setting, arises from the misty intervening particles, which reflect the light from these bodies without decomposing it.

The tides at the new and full moons must have some influence upon the weather, particularly when it is sultry in summer. Thus, it is cooled by the under water being forced up from their swelling and action; by their passing up rivers and causing them to rise higher and to cover the low grounds on their margins; and likewise, by rendering them of a lower temperature from the dilution of the salt sea-water. This is further diminished by the tides covering the sands near the coasts, and by mingling the colder fluid over shallows with that near the surface. In consequence of the above causes, the evaporation during the night is greatly lessened, and the aqueous vapour above much condensed; by which means, when there is little wind, clouds are prevented from forming, and the radiation from the surface is considerably increased, and its temperature greatly reduced. When the moon is about full, the superior clearness of the nights may, in great part, be owing to the above circumstances.

THE SENSATIONS OF ANIMALS.

With respect to animals, they are observed to become uneasy and restless before rain, and whole pages of many writings on meteorology, from Aratus downwards, are filled with prognostications from the motions or cries of different birds, beasts, fish, reptiles, and insects. It may be sufficient to give a few instances : thus, swallows fly low—in pursuit of the insects they feed on, seeking the warmer air in the change of weather coming on; ducks, geese, and other aquatic birds are more noisy and active than usual, because some of these like wet weather, and others must look to the land for food in the coming disturbance of the ocean; spiders " put their house in order" by drawing in their tackle—" shutting up" completely; worms rise in greater numbers; and leeches kept in glasses are observed to move about frequently—before rain.

Human beings are affected in various ways previous to stormy weather,—in the places of wounds that have been long healed, in rheumatic pains, and in those parts of the system which are delicate or diseased.

These indications are much stronger in summer than in winter, probably owing to the quantity of aqueous vapour in the atmosphere, which must influence the passing off of moisture by the breath from the lungs, or by the perspiration of the body. In hot weather, the greater or less expenditure of aqueous fluid from the system must become more considerable, and attended with increased influence.

That the effect does not depend upon the greater or less pressure of the atmosphere is evident from the ascent in balloons, or the climbing of mountains, not producing the same sensations.

Some meteorologists, however, insist on this cause—the diminished pressure of the atmosphere, as shown by the barometer, whence there results an immediate increase in the rise of the gases in the interior of bodies. Thus they explain the rising of leeches in bottles, or of small frogs similarly confined, which, in certain countries, are furnished with a small graduated scale, to serve as barometers. The same cause produces the disturbance of marshes, etc., and permits bubbles of gas and the weeds in them to rise to the surface. It is the same with the gases of cesspools, sewers, etc., whose smell is then most offensive. But then, we know that moist air increases the transmission of odours by the air, so that the rise of the matters is one thing, and their increased smell another. The springs of gaseous waters, not thermal, bubble up more on such occasions, and the taste of those which contain carbonic or other acid is diminished, whilst the quantity of water delivered by certain fountains is augmented.

There seems in all this a combination of atmospheric pressure and atmospheric electricity. According to

Peltier, when a storm is on the point of forming, that is when the air is saturated with water charged with a powerful negative electric tension, the influence of the latter repels the negative electricity of the soil, decomposes the natural electricity, and attracts the positive electricity to the surface.   Living bodies, especially mankind, are then, in this case, in an electrical state completely opposite to the electrical state which is natural to them.   Hence, he says, the state of uneasiness which prognosticates stormy weather.   When several electric discharges have taken place, especially when rain has begun to fall copiously, this uneasiness ceases rapidly. The reason is simple; the electric discharges have diminished the negative tension of the storm—so has the rain—partly because each drop carries off a certain quantity of electricity, and partly because the air situated between the storm and the ground becomes moister, and consequently a better conductor.   Then all living bodies, men especially, become less " vitreous" or " positive," and the state of uneasiness and discomfort diminishes.

Saline or salt impregnations, from imbibing moisture, are remarkable indicators of approaching rainy weather. When they deliquesce, from their surfaces being colder than the external air, it is owing to their condensing the aqueous vapour which passes rapidly in advance of the wind through the atmosphere.

Soot frequently falling down chimneys is an indication of coming rain; the aqueous vapour, increasing before rain, is imbibed by the carbonaceous particles of the soot, which falls down by mere increase of weight.

The more rapid the heating of fires, etc., the drier and denser the atmosphere may be considered, and fair weather prognosticated. When fires burn less vividly, the air is damp, and rain may be expected.

Smoke falling indicates rain ; either it imbibes moisture from the saturated air, or the decreased density of the air cannot permit it to ascend.

CHAPTER XVIII.

THE INFLUENCE OF THE WEATHER ON WOOD, AND HOW TO PREVENT IT.

In the celebrated poem on Weather Signs[1] ascribed to Dr. Jenner, but to Darwin by Forster, we find the following—" Hark! how the chairs and tables crack." This is incorrect, referring to the signs of *rain*. The cracking of wood indicates fine weather, because it is the result of

[1] The *hollow winds* begin to blow;
The *clouds look black*, the *glass is low;*
The *soot falls down*, the *spaniels sleep;*
And *spiders* from their *cobwebs peep.*
Last night the *sun* went *pale to bed;*
The *moon* in *haloes* hid her head.
The boding shepherd heaves a sigh,
For, see, a *rainbow* spans the sky.
The *walls are damp*, the *ditches smell,*
Closed is the pink-eyed *pimpernel.*
Hark! how the *chairs* and *tables crack,*
Old Betty's joints are on the rack :
Her *corns* with *shooting pains* torment her,
And to her bed untimely sent her.
Loud *quack the ducks*, the *sea-fowl cry,*
The *distant hills* are *looking nigh.*
How restless are the *snorting swine !*
The *busy flies* disturb the *kine.*
Low o'er the *grass* the *swallow wings,*

dryness in the air. On the contrary, it is the swelling of
wood which makes our doors and windows hard to shut in
rainy weather. Indeed, this property of wood has been
turned to account in the manufacture of millstones. A
block of stone is divided into circles by holes pierced in
it, which are filled with wood dried in an oven. The
stone is then exposed to the air in moist weather, when

> " The *cricket*, too, how *sharp he sings !*
> *Puss* on the hearth, with *velvet paws*,
> Sits *wiping* o'er her *whisker'd jaws*.
> The *smoke* from *chimneys right ascends*,
> Then, spreading, *back to earth it bends*.
> The *wind* unsteady *veers around*,
> Or settling in the *south is found*.
> Through the clear stream the *fishes rise*,
> And *nimbly catch* the incautious *flies*.
> The *glowworms* num'rous, clear and bright,
> *Illumed* the *dewy hill* last night.
> At dusk the squalid *toad* was seen,
> Like *quadruped*, stalk o'er the green.
> The *whirling wind* the dust obeys,
> And in the *rapid eddy* plays.
> The *frog* has changed his *yellow vest*,
> And in a *russet coat* is drest.
> The *sky is green*, the air is still,
> The *mellow* blackbird's voice is shrill.
> The *dog*, so alter'd in his taste,
> Quits mutton-bones, on *grass* to feast.
> Behold the *rooks*, how odd their flight,
> They imitate the *gliding kite*.
> And seem *precipitate to fall*,
> As if they felt the piercing ball.
> The tender colts on back do lie,
> Nor heed the trav'ller passing by.
> In *fiery red* the *sun* doth *rise*,
> Then wades *through clouds* to mount the skies.
> 'T will *surely rain*, we see't with sorrow,
> *No working in the fields to-morrow.*

the wood swells to such a degree as to split the stone as effectually as though by iron-wedges driven in by sledge-hammers.

The swelling of the wood is a consequence of the entrance of moisture into the fine tubes where the original sap before circulated, and its shrinking ensues upon the expulsion of the liquid from these tubes by the elastic force of the woody fibre tending to resume its former state. The attraction between water and the woody matter undoubtedly affects the surface, and is concerned in the rotting of the substance of the wood. It follows as a consequence that to secure wood panels from swelling by damp and cracking afterwards, and doors from being " set fast," it is not enough to cover the outer surface, or even both surfaces, with a timely coating of paint; the ends and sides of the panels, where they are let into grooves, should above all be attended to and well pitched or painted—if we do not adopt carbonization—in order to prevent the entrance that way of water, which will otherwise infallibly ensue and convert our woodwork into a perpetual hygrometer—useful, no doubt, as an indicator of coming weather, but by no means desirable in the matter of comfort and convenience. There seems actually to be a real though feeble chemical affinity between the fibre of wood and water, whereby the wood will decompose the vapour which constantly exists in greater or less quantity in the atmosphere and appropriate its water. The attraction between the woody matter undoubtedly affects the surface and is concerned in the rotting of the substance of the wood, but it has little to do with this peculiar process, which is wholly the result of the vegetable organization, and continues only while the wood remains in some measure elastic.

When this elasticity ceases, after some years, by the beginning of a change tending to decay, the wood is pro-
nounced to be "seasoned," and is no longer found to expand and contract as before; but, evidently, the principle of decay is still in it; no extent of seasoning can
kill it; such is the natural process of decay in wood.

The preservation of wood, whether in our buildings on
land or in our ships in the water, on our rivers and the
ocean, is of the utmost importance, and any effective
method that can be suggested deserves consideration.
This requirement is rendered more imperative by another
consideration, namely, that the continuous clearance of
forests in all countries and the increasing demand for
wood of all kinds seem to threaten a coming time of
scarcity of the material. Immense quantities are used
in every country, in spite of the favour shown to iron of
late years. France and England use annually each at
least some ten million cubic yards of wood. There will
be no possibility of ensuring the requisite supply by the
natural process of replanting at this rate of consumption.
and therefore it is time to think of directly diminishing
the waste of the precious material by means of its preservation. Various modes of preserving wood have
been suggested and applied, but all of them are costly
and complicated. The fact is, in this matter we have
been trying to do in a roundabout way what has been
readily and directly done from time immemorial. Every
farm labourer knows the primitive mode of preserving
wood from decay. Before fixing a post in the ground
he burns the foot of it in a heap of chips or brambles,
and the consequence is that when exposed to the moisture of the soil, which would cause its decay, this part
of the post remains perfectly sound when the remainder

shows the incipient signs of rot and decomposition. Carbonization is, therefore, a powerful means of preservation, and its effect is easily explained. In the first place, heat, by drying the external parts of the wood, renders them inaccessible to fermentation or the natural principle of decay in the wood itself. In the second place, burning hardens the layer immediately below the carbonized portion; and this layer, thus baked, as it were, and consequently impregnated with empyreumatic essences (whose antiseptic properties are well known), wonderfully resists the influence of the humidity of the soil and atmosphere. It is utterly impossible that any mouldiness or fungus of any kind can form on a layer of carbon or a carbonized surface, and mouldiness and fungus are the sure precursors of corrupting fermentation.

Thus carbonization has not only been found practically sufficient for the preservation of wood, but it is proved to be so on scientific principles. Strange that the process has not been thought of by our painters, who must have repeatedly observed that after having to scorch off old paint from doors, etc., there has been invariably less decay and shrinking in it. Pianoforte-makers also have found the advantage of baking their wood, in order to deprive it of its natural property. In fact, carbonization is the only infallible means of depriving wood of its hygrometric property, or tendency to be affected by changes of the weather as to moisture or dryness, which is the everlasting cause of the cracks and swelling universally complained of. Mere painting is of little avail to protect wood from the effects of moisture in the air whilst the fermenting principle in it remains unchecked. At any rate, in mortising, all the joints of constructions should be well served with paint or pitch, if we persist in trusting to this method alone.

The carbonization of the wood is the only remedy, which acts not only by closing the pores of the wood to moisture, but also by arresting its fermentation.

The only difficulty is to apply the process to pieces or assemblages of worked-up materials without injuring the joints or causing the wood to shrink. Now, this difficulty has been overcome by M. de Lapparent, the Director of Naval Constructions in France. As far back as 1862, M. de Lapparent suggested the carbonization of the hulls of ships as the most efficacious remedy for the rapid decay of these wooden carcases, owing to the moisture caused by the introduction of steam engines. He proposed to effect the carbonization by means of a jet of flame directed and applied by compressed air. This is now effected, the apparatus required being extremely simple. Two tubes of india-rubber—one leading from a reservoir of ordinary gas, the other from bellows worked by means of a treadle—convey into a copper tube simultaneously both the gas and the compressed air. As soon as the mixture is inflamed we obtain a flame sufficient to solder metals. Keeping the bellows in action with the foot, all we have to do is to direct the mouth of the copper tube with one hand against the surface of the wood in position, when it will be carbonized rapidly and uniformly throughout its entire extent. The method has been adopted in the French arsenals, where entire ships are thus carbonized, the different parts of the hull being subjected to this operation as soon as they are placed in position. The utility of the process is unquestionable; but its general application presented, until recently, very great difficulties, because lighting gas is

only to be procured in large towns, and even in these it
is only at night that we can get adequate pressure. The
problem, therefore, still remained to be solved for agricultural and private purposes. The son of the original
inventor has overcome the difficulty, and contrived an
easy substitute. This consists in a portable lamp adapted
for general use in the process, by means of common tar-
oil or paraffin. The oil ascends by the capillarity of a
cylindrical wick of large diameter, placed horizontally at
the side of the reservoir; in the centre of the wick is a
pipe which communicates with a pair of bellows worked
by a treadle; a metallic chimney, pierced with holes at
its base, completes the apparatus, which gives a flame as
intense as a jet of gas. Being very portable, this appliance enables an operator to carbonize, in any situation,
about three square yards of oak per hour, and eight
square yards of poplar or deal in the same time, with a
very small expenditure of oil.

A recent occurrence suggests that the spars, etc., of
ships, should be subjected to this process of carbonization. During the late cruise of the Channel squadron,[1]
the 'Ocean' sprang her topmast crosstrees, and had to
send down her maintopmast,—when examination showed
that both trusseltrees were rotten. As this ship is perfectly new, it is evident that the wood of her spars and
crosstrees was in a state of decay before being used for
the purpose. All wood kept in store should therefore be
carbonized. The French railway companies have availed
themselves of the discovery, applying it to all their
constructions and rolling-stock. To farmers and all
concerned in agricultural erections, the discovery is invaluable, considering its application to malting-houses,

[1] October, 1866.

cow-sheds, etc., which are continually exposed to a warm and moist atmosphere, which the respiration and exhalations of the animals still further impregnate with organic corpuscles—the primary cause of fermentation. In like manner, it is well known that the wood composing the parts of agricultural implements, which are daily exposed to the alternations of moisture and dryness, often becomes unserviceable before the instruments themselves are used up.   Herein, undoubtedly, a previous carbonization would prove a most efficacious and economical remedy.   Finally, this process will be found eminently adapted for the preservation of enclosures, gates, railings, and especially hop-poles, the costly maintenance of which would thereby be reduced by two-thirds at least.[1]

[1] The above appeared originally in the 'Mechanics' Magazine ;' and in the number for Nov. 23, 1866, will be found the description (also communicated by the Author) of another method of carbonization, with figures, invented by M. Hugon, which is also used for blasting purposes instead of gunpowder, etc.

CHAPTER XIX.

THE SUPPOSED INFLUENCE OF THE SAINTS ON THE WEATHER; THE PERIODICITY OF CHOLERA AND ITS ATMOSPHERIC CAUSE.

THE popular proverbs and rhymes of all nations on the weather may be considered the groundwork of the learned science of Meteorology. Nay more,—our modern science has in almost every instance confirmed, whilst explaining, those rude hints of popular experience, showing that they were founded on the laws of nature as ordained by Providence. Our Saviour alludes to two of these popular sayings, as recorded by St. Matthew, xvi. 2, " When it is evening ye say, It will be fair weather, for the sky is red." Modern science explains this adage. The colours of the sky result from the state of the watery vapour in it, and the various tints depend upon the condition of that vapour. The cause of a *rosy* sunset, alluded to in the proverb quoted, is that the vapour of the air is not actually condensed into clouds, but only on the point of being condensed; in which state it bends the *red* rays of the sun towards the horizon, where they tint the floating clouds; and the reason why a rosy sunset is an indication of a fine day to follow, is because, notwithstanding the cold of sunset, the vapours of the earth are not condensed into *clouds,* and therefore there can be no rain, for which, in general, clouds must be formed.

Again, our Saviour says :—" In the morning ye say, It will be foul weather to-day, for the sky is red and lowering." (Matt. xvi. 3.) The cause of a red sunrise is that the vapour in the air is just on the point of being condensed; the higher regions of the air are laden with vapour, on the very point of condensation, which the rising sun cannot disperse, and therefore it must soon fall in showers.

Numerous popular maxims about the weather are connected with our saints' days throughout the year. Thus we find the old verses respecting Candlemas Day :—

> " Si Sol splendescat Mariâ purificante,
> Major erit glacies quàm fuit antè."

Which is paraphrased as follows in the Scottish rhyme : —

> " If Candlemas Day be dry and fair,
> The half o' winter's to come, and mair ;
> If Candlemas Day be wet and foul,
> The half o' winter 's gane at Yule."

St. Paul's Day, 25th January, is another weather-period :—

> " Clara dies Pauli tempora denotat anni bona ;
> Si nix vel pluvia, designat tempora cara ;
> Si fiant nebulæ, pereunt animalia quæque ;
> Si fiant venti, designat prælia genti."

> " If St. Paul's day be fair and clear,
> It does betide a happy year ;
> But if it chance to snow or rain,
> Then will be dear all kind of grain ;

If clouds or mists do dark the skie
Great store of birds and beasts shall die;
And if the winds do flie aloft
Then war shall vex the kingdom oft."

With respect to the value of these two weather-maxims, we may say that the first, referring to the 2nd of February, may be to a certain extent meteorological. Dry and fine weather at a time when the reverse is usual, denotes an atmospheric disturbance likely to follow, as announced in the adage; but the comprehensive results bespoken by the various "aspects" for St. Paul's Day, seem to warrant it a mere astrological origin.

Of course something must be said of St. Swithin's Day. As every one knows, if it rains or is fair on St. Swithin's Day, the 15th of July, there will be a continuance of wet or dry weather for the forty days ensuing.

"St. Swithin's Day, if thou dost rain,
For forty days it will remain;
St. Swithin's Day, if thou be fair,
For forty days 'twill rain nae mair."

Authoritative as seems the proverb, it is, however, quite certain that there is little or no reliance to be placed in it. From observations made at Greenwich for the twenty years preceding 1861, we find that the greatest number of rainy days after St. Swithin's Day, had occurred when the 15th of July—the day in question —was *dry*.

But although the precise day has no meteorological significance, the popular notion is not altogether unfounded if it be referred to the *period* of the year. It is, indeed, likely enough that a track of wet weather, or the opposite, may occur at this period of the year, because a change generally takes place soon after Midsummer, the character of which will depend much on the state of the previous spring. If this has been for the greater part dry, it is very probable that the weather may change to wet about the middle of July, and *vice versâ*. So we must interpret this proverb with reference to the preceding spring, and then, perhaps, we shall not be far out in our calculation. The last St. Swithin's Day was certainly "fair," but we have not had the promised forty days' continuance of fine weather. The previous weather has been, however, much of a character to correspond with the inference from that of the previous spring.

We may observe that there are other rainy saints'-days besides St. Swithin's. Thus in France they have the *St. Medard* (June 8), and the *Saints Gervais and Protais* (June 19), with a similar forty days' rain ascribed to them.

The origin of St. Medard's Day is declared as follows, in a very pretty legend :—Medard being out with a large party one hot summer day, a heavy fall of rain suddenly took place, by which all were thoroughly drenched, with the exception of the saint himself, round whose head an eagle kept continually fluttering, and, by sheltering him with its wings till his return home, accomplished effectually the purpose of an umbrella. We need scarcely remind the reader that according to the old legend, St. Swithin's rain of forty days was a punishment for the disturbance of his remains, contrary to his expressed wish on the subject.

St. Vincent's Day (22nd January) is of greater meteorological importance. The adage respecting this day is rather obscurely worded :—

x

> " Vincenti Festo
> Si Sol radiet,
> Memor esto."

That is, "Should the sun shine bright on St. Vincent's Day, remember it." The matter was a complete mystery to modern investigators of folklore, until a gentleman residing in Guernsey, looking through some family documents of the sixteenth century, found a scrap of verses, expressed in old provincial French, as follows :—

> "Prends garde au jour St. Vincent ;
> Car, sy ce jour tu vois et sens
> Que le soleil soiet cler et beau,
> Nous erons [aurons] du vin plus que l'eau."

That is, "Should the sun shine clear and bright on St. Vincent's Day, we shall have more wine than water." It is clearly, therefore, a hint to the *vine-culturing peasantry* that the year would be a dry one and favourable to the vintage ; and the maxim was doubtless founded on the observation that fine weather on that day was generally followed by fine weather for the season.

Among these popular adages consecrating certain dates of the year to particular weather, there are the *Saints de Glace* (*icy saints*) and the *Été de St. Martin* (St. Martin's summer) universally noted in France, and to which recent meteorological researches have given some importance, in connection with astronomical causes.

> " Saint Mamert, Saint Pancrace
> Et Saint Servais—
> Sans froid ces saints de glace
> Ne vont jamais."

Such is the agricultural proverb which announces for

the 11th, 12th, and 13th of May—the anniversaries of these saints—a notable refrigeration in the mean temperature at that period. The temperature rises rapidly during the last day of May, but the French farmers have for a long time remarked, and regularly-kept meteorological observations attest, that during the first fortnight of the month there occurs a notable refrigeration. Hence the supposed influence of the "icy saints," St. Mamert, St. Pancrace, and St. Servais. The temperature, after rising regularly every day during April and sometimes the first days of May, suddenly suffers a remarkable lowering, often accompanied with frost. Until that critical period is passed, the harvest is in danger. All the cereals may then be struck down by sudden frost.

On the other hand, the 11th of November, St. Martin's Day, or rather the preceding and subsequent days, are, on the contrary, remarkable for a notable increase in the mean temperature, constituting the second or "Indian summer" of France.

Here, then, are two anomalies, the observation of which is attested both by a long tradition of the rural population and by the investigations of modern meteorologists. Doubtless, in the course of the year, we may experience analogous perturbations; but these are oftener mere local variations, irregular in their dates, and consequently they do not present that periodicity which characterizes the refrigeration of the first days of May, and the high temperature of the first fortnight of November. Moreover, whilst most rural proverbs are of double meaning and conditional, those relating to the "icy saints" and St. Martin's summer are affirmative, which is very rare.

There is another adage relative to February 6th :—

"Quand vient la Sainte Dorothée,
La neige est plus épaisse."

This refers to the 6th of February. The 10th of August, or *Saint Laurent,* is also a critical date for agriculturists ; but like that of the 9th of February it is less remarked and passes off, so to speak, unperceived. This difference is easily explained. When the temperature is rapidly increasing, as in May, a sudden refrigeration is perceived by everybody ; it is the same in November, for the same reason. But a momentary refrigeration in the midst of a cold season, as in February, and an increase of temperature during the great heats of summer, can only be remarked by those who attentively check meteorological variations.

Thus we have periodical perturbations of temperature in May, November, February, and August. The perturbation of May is general, from the south to the north of the temperate zone.

And now for the scientific explanation of these anomalies. Professor Erman, of Berlin, writing to the celebrated French astronomer Arago, in 1840, gave the following opinion :—" The two swarms or currents of planetary bodies [meteors, shooting stars, etc.] which the earth meets on the ecliptic, respectively about the 10th of August and about the 13th of November, annually interpose themselves between her and the sun,—the first during the days comprised between the 5th and the 11th of February, the second from the 10th to the 13th of May. Each of these conjunctions causes annually, at these periods, a very notable extinction of the calorific rays of the sun, and thereby lowers the temperature at all the points of the earth's surface."

Let us pause for a moment on this theory of Professor Erman, and see how it can account for the refrigeration of the months of February and May, and confirm the popular sayings relative to the icy saints and the anniversary of St. Dorothea. What connection is there between the two phenomena in question and the apparition of numerous clusters of meteors or shooting stars furrowing the skies on the nights of the 10th of August and the 13th of November?

Amongst the various hypotheses advanced to account for shooting stars, the most probable is that they form parts of a ring, or rather of a series of rings of small planetary bodies revolving round the sun, at about the same distance as the earth from that central body. Now, the earth, in her annual orbit or path round the sun, passes by those innumerable banks, as it were, of miniature planets; but, as the plane of her orbit does not appear to coincide exactly with the planes of the meteoric rings, the consequence is, that our earth is not always in those regions of the heavens which are equally thronged by these singular planetary bodies. About the 10th of August, however, we cross one of these banks—right through its diameter; hence, that multitude of luminous meteors, shooting stars, falling stones, etc., furrowing the nightly skies with their flaming trajectories. It is evident that it is only the brightness of daylight that prevents us from observing this phenomenon in the daytime, since the apparitions take place with reference to all parts of the earth at the various moments of the same daily rotation. Brushing through our atmosphere, these small bodies, constituted for the most part of inflammable substances, blaze for an instant in sight, and then continue far away their celestial course through space.

Some of them penetrate more deeply into the sphere of the earth's attraction, and being thus drawn out of their course, they fall over our heads and burst into fragments upon the soil, thus presenting us with the most curious specimens of the matter of which the other planets are composed. Six months later, that is, in February, the same swarm a second time cuts the plane of the earth's orbit; but then, as their distance from the sun has doubtless varied on account of the elongated form of the planetary orbits, we can no longer see them, because the ring happens to be between our earth and the sun.

It is this interposition—this sort of eclipse of the sun caused by the multitude of small planetary bodies, rendered invisible by their small dimensions—that, according to Professor Erman, accounts for the lowering of the temperature of February. In like manner, a second swarm or ring is crossed by the earth in November, and interposed between the latter and the sun, six months later, namely, in May. These are the icy saints, *St. Mamert, St. Pancrace,* and *St. Servais,* metamorphosed by science into troops of small satellites revolving round the sun.

But if the refrigeration be caused by a sort of eclipse of the sun effected by these small planetary bodies, how are we to account for their producing an increase of temperature in August and November? The reply is,—that whilst in February and May they merely intercept a portion of the heat radiated to the earth by the sun, they, on the contrary, in August and November, as solid bodies, check and diminish the earth's radiation in space, and so retard her refrigeration, by returning to her a portion of the heat which they themselves receive from the sun.

Of the great fact itself of these periodical perturbations there is no doubt whatever; the only difficulty is to account for them; and the above hypothesis, strange as it appears, if not entirely satisfactory, is supported by very eminent meteorologists, and has been most elaborately worked out, through the examination of the meteorological records of nearly 150 years, by M. Sainte-Claire Deville, who has awakened a new interest in these mysterious meteors or shooting stars. For ages they have been considered prognostics of the weather, and M. Coulvier Gravier, who has devoted many years to the study of them, believes that they show the direction of the coming wind; that their slow motion foretells a calm, to ensue or to continue if it exists; in fact, he says they are our weathercocks or anemometers in the upper regions of the sky. But M. Deville draws another and much more important conclusion from the theory which we have explained,—namely that these periodical perturbations, in their *maxima*, may be connected with the health of the human race and the lower animals, and that of the vegetable creation. He asks:—" Do not all these considerations almost necessarily lead us to infer the influence of these critical periods, by their sudden variations of temperature, not only on the health of the vegetable creation, but that of the human race? Should we not examine the registers of hospitals to see if certain diseases are not more frequent on certain *days* of certain *years?* Can we not even go back to the past and see, in the history and the chronicles of past ages, if there was some trace of *periodicity* in certain great perturbations in the health of nations,—like the two invasions of *Cholera,* which, perhaps by chance, occurred in 1832 and 1849, about the *centre* of the two critical periods, and which came from the *North*, like the *Aurora Borealis*,—since it seems also that it is these great atmospheric waves that propagate the perturbations of temperature?"

Now, it is at least most remarkable that the same interval of *seventeen* years should have occurred between the visitation of cholera in 1849 and the recent attack in 1866, as occurred between 1849 and 1832!

At any rate, there seems good ground for believing that we have at length got a clue to the *periodicity* of this dreadful plague; and if so, the advantage we may take of it in the future, will be attended with immense results in enabling us to prepare for the visitation.

CHAPTER XX.

THE CURIOSITIES OF LIGHTNING.

THE circumstances which increase atmospherical electricity are, regular thunderclouds; a driving fog and small rain; snow or brisk hail; a shower in a hot day; hot weather after wet, and wet after dry; clear weather, hot or frosty; a cloudy sky; a mottled sky; sultry and hazy weather; a cold damp night; southerly winds.

All nature is subject to this excitement, of which little is known beyond its effects. The human body and all animal bodies are electrical or galvanic combinations, and the excitement is the principle of vitality and its consequent energy. The surfaces oppositely excited are those of the lungs and the skin. The lungs fix oxygen and are positive; while the skin fixes an equivalent and is negative. The circulations, secretions, etc., are intermediate results, and the action of the heart arises from the proximity of the positive arterial blood to the negative venous blood. The action exhausts itself, as it ought, in the system. In 2422 observations different persons were 252 times externally positive, 771 were negative, and 339 imperceptible. In sitting at rest Hemmir found himself 332 times positive, 14 negative, and 10 times imperceptible. Rest and action produce changes, owing to the varied effect of the lungs and skin, and the nervous system appears to act by a similar movement.

Respiration renders the air of rooms negative when persons are at rest, owing to the lungs being in action while the skin is quiescent. School-rooms and sleeping-rooms become negative, while the external air is positive.

These facts lead to useful conclusions. Thus, the reason why closed windows are dangerous during light-ning is, because the inner side of the panes acquires an opposite electricity to the outside, and then any con-ducting body is likely to concentrate the action on the inside. Still, all draughts of air are conductors, and should be shunned. Metallic bodies, picture-frames, coated mirrors, bell-wires, etc., display electricity by in-duction during a storm.

The best lightning conductor is said to be lead or copper on the ridge of the roof, with perfect continuation of metal pipes into the ground.

The frequency of disastrous thunderstorms of late years, during the present especially, naturally directs more than a usual amount of public attention to these formidable phenomena of nature. The sacrifice of hu-man life with which these visitations have been accom-panied, and the narrowness of the escape which many persons have experienced, induce a certain degree of general apprehension.

Of course in all this there is nothing new, and yet it would seem that electric visitations move somewhat in cycles,—that for a few years there may be a comparative immunity from deaths by lightning, after which a season or two may be noted for fatal displays of the electric force.

The recent catastrophes all over the country suggest the inquiry whether anything more can be done to protect mankind against the perils of lightning. The use of the metallic conductor for the protection of ships and buildings, is now tolerably common. Perhaps, however, this source of safety is not sufficiently resorted to. Spires, columns, and public buildings are, generally speaking, guarded in this way. But there are many private residences which need this kind of protection as much as Government establishments and other public edifices. Several dwelling-houses have lately been struck with lightning, and it is not unreasonable to suppose that the event was due to a lack of some other channel for the escape of the electric fluid. We do not mean that every house should have a lightning conductor, for we believe this would be dangerous; but if metallic conductors were provided at reasonable intervals throughout the country, and in towns and cities, it is just possible that the actual severity of our storms would be mitigated, and the loss of life both out of doors and within, considerably diminished. Lest our readers should smile at the idea, we may just mention the fact that the vineyards of France, Switzerland, Germany, and Italy, are said to have materially benefited by the erection of electric conductors at distances of about 120 feet from each other, the effect being to prevent the occurrence of hailstorms. At the close of the vintage, the poles or *paragrèles* (protectors from hailstones) are put away, and at the vernal equinox they are brought out and stuck up again. The powers of nature are more at our disposal than we are apt to believe—if we only study and take the numerous hints she gives us all around creation.

The statistics of lightning strokes present some curious facts. From a note presented to the French *Académie des Sciences*, by M. Boudin, it appears that during the period elapsed between 1835 and 1863, that is, twenty-nine years, no less than 2238 persons were struck dead by lightning. The annual maximum has been 111, the minimum 48; but if we add the number of the injured to that of the dead, the total number of the victims of lightning exceeds 6700, and the average per annum is 230.

It appears that females are much less liable to be struck by lightning than males; thus, of 880 victims struck between 1854 to 1863, there were only 233, less than one-third, of the former.

It is impossible to account for this immunity of the weaker sex, unless it be ascribed to their silk garments, —but the fact is incontestable; in many cases the lightning, in falling on a group of persons of both sexes, has picked out the males.

Another very curious fact appears from these statistics, namely, that some persons seem to be more liable to be struck than others. Some have been struck by lightning several times during their lives; and what is stranger still, one of them has been struck three times by lightning, in three different dwellings! Surely this man should have taken the hint to provide himself with a perpetual lightning-rod and conductor, from the vernal to the autumnal equinox!

Out of 6714 persons struck by lightning, it is not surprising to find that about one-fourth have been struck *under trees*, in spite of the everlasting warning as to the avoidance of that fatal neighbourhood during thunderstorms. Obviously people rush under trees to get shelter from the rain; but, when the terrible danger is known we apprehend that few will do so. It is quite certain, from the present statistics, that 1700 persons out of the 6714, would have escaped death or severe wounds by avoiding the vicinity of trees during the thunderstorms.

In accordance with what we have previously written, the highest localities suffer most, and the plains least, from thunderstorms.

The electric condition of objects struck by lightning is equally curious. It seems, according to many observers, that objects struck by lightning remain "charged" with electricity, like the Leyden jar of experiment, and retain 'for a time the faculty of striking any other object that may approach them. Two cases of the kind were instanced by M. Boudin. On the 30th of June, 1854, a man was killed by lightning near the *Jardin des Plantes*, at Paris, and his body remained for some time exposed to a pouring rain. After the storm two soldiers, on attempting to lift the corpse, received a shock the moment they touched it. On the 8th of September, 1858, two artillerymen, on attempting to raise two telegraphic poles which had been struck down by lightning at Zara, in Dalmatia, having laid hold of the conducting wire, two hours after the storm, first felt slight shocks, and then were instantly struck down. Both of them had their hands burned, and one of them seemed totally lifeless. The other, on endeavouring to rise, fell down immediately on touching with his elbow one of his comrades who ran up at his cries for assistance. The latter was also struck down, his nervous system was seriously shaken, and his arm exhibited a burn on the skin at the spot were he was touched by the man struck by the lightning.

Like every electric spark, lightning always follows the

best conductors, chiefly the metals; but still it will leave
metal for a body which, although not so good a conductor, conducts it more directly to the earth. Next to
the metals, damp substances are followed by it in preference: hence, men and animals are often struck by it.
Death appears to be caused by a shock on the nervous
system; for dead persons retain the very same position
which they occupied before they were struck by lightning.[1] The following sad case is one of the most interesting on record:—

M. Buckwalder, a Swiss engineer, had established a
geodesical or surveying signal on the top of the Sentis
mountain, in the canton of Appenzell. This summit is
about 2504 yards above the level of the sea. "On the
4th of July, 1832," he writes, "it rained abundantly
towards evening, and the cold and wind became such
that they prevented my sleeping all night. At four
o'clock in the morning, the mountain was covered with
clouds, and some passed over our heads; the wind was
very violent. However, larger clouds coming from the
west, approached and slowly condensed. At six o'clock
the rain began again, and the thunder resounded in the
distance. Soon the most impetuous wind announced a
tempest. Hail fell in such abundance that, in a few
moments, it covered the Sentis with a frozen stratum
which was an inch and a half in thickness. After these
preliminaries the storm appeared calmer; but it was a
silence, a repose during which Nature was preparing a
terrible crisis. At a quarter past eight o'clock, the
thunder growled again; and its noise, approaching
nearer and nearer, was heard without interruption till
ten o'clock. I went out to examine the sky, and to

[1] Kaemst, *ubi suprà.*

measure the depth of the snow at a few paces from the tent. Scarcely had I taken this measure, when the lightning burst forth with fury, and obliged me to take refuge in my tent, together with my assistant, who brought some food there to take his repast. We both lay down side by side on a plank.

" A thick cloud, as dark as night, then enveloped the Sentis; the rain and hail fell in torrents; the wind blew with fury; the near and confused lightnings seemed like a conflagration. The lightning broken into flashes mixed its horrid bolts, which, driving against each other and against the sides of the mountains, indefinitely repeated in space, were at once an acute rending, a distant reverberation, and a deep and long roaring.

" I felt that we were in the very centre of the storm; and the lightning showed me this scene in all its beauty or in all its horror. My assistant was overwhelmed with fear, and he asked me if we were not in danger. I removed his fears by relating to him that, at the time when MM. Biot and Arago were making their geodesical experiments in Spain, the lightning had fallen on their tent, but had only passed over the roof without touching them. As for myself, I was really at ease; for, accustomed to the noise of thunder, I still studied it when it threatened me closely. These words, however, brought to my mind the idea of danger, and I fully understood it. At this moment a globe of fire appeared at the feet of my companion, and I felt my right leg struck with a violent commotion—which was an electric shock. He uttered a doleful cry : ' Ah ! my God !' I turned round to him. I saw on his face the effect of the lightning-stroke. The left side of his face was covered with brown or reddish spots. His hair, eyebrows, and eyelashes were frizzled and burned; his lips and nostrils were of a brownish violet; his chest seemed still to heave at intervals; but soon the sound of respiration ceased.

"I felt all the horrors of my situation; but I forgot my sufferings in order to seek succour for a man whom I saw dying. I called him, but he did not reply. His right eye was open and bright; it seemed to me as though a ray of intelligence beamed from it, and I hoped: but the left eye remained closed; and, on raising the eyelid, I saw that it was dull. I supposed, however, that there was still sight remaining on the right side, for I endeavoured to close the eye on that side,—an attempt which I repeated three times. It opened again of itself, and seemed animated. I put my hand on his heart. It beat no longer. I pricked his limbs, body, and lips with a compass: all was immovable. It was death;—and I could not believe it!

"Bodily pain at last drew me from this painful contemplation. My left leg was paralysed, and I felt a shuddering, an extraordinary movement. I felt, besides, a general trembling and oppression, and disordered beatings of the heart. The most sinister reflections took possession of me. Was I going to perish like my unfortunate companion? I thought so, from my suffering; however, reason told me that the danger was past. I gained with the greatest difficulty the village of Alt St. Johann. The instruments had been struck in like manner."[1]

Among the most recent instances may be mentioned that of the 30th of June, 1866, at Wormwood Scrubbs,

---

[1] 'Ergebnisse der Trigonometrischen Vermessungen in der Schweiz,' and *apud* Kaemst, *ubi suprà*.

when a boy was killed by the lightning, and two of his brothers and his father were struck down. The father said: "A flash of lightning came and struck us all down. It struck my three sons and myself, and also a retriever dog. I first crawled to my eldest son, and found him quite insensible. I thought he was dying or dead. I then went to my other son, and found him with his hat cut up. He was lying on his face. I turned his head round, and saw he was dead. The lightning struck his head, tearing his cap, and went down under his clothes, tearing his left boot. It also killed the dog. It did not burn his clothes, except his neckerchief." Another witness said: "The weather had got somewhat clearer, when, without the slightest warning, a terrific crash of thunder came. I saw fire come down, and then a vapour rose up, preventing my seeing anything for a time. When that cleared off, the last witness, his two sons, and the other man were lying on the ground. I was myself struck by something which I took to be spent shot. One person complained of being touched on the side of the face,—indeed, several persons felt something strike them. I noticed the dog was kicking its hind legs when the vapour cleared up." According to another witness, the deceased's hair was singed at the right side of the head; blood flowed from the mouth and nose when the head was moved; the face and hands were discoloured. All but the deceased completely recovered.

Near the spot where lightning has fallen, an odour is perceived similar to that which strikes the nostrils in the neighbourhood of electrical machines when in action. It has always been said that this is a sulphurous odour; but we must not forget that the mass of mankind designate by this term every disagreeable odour which cannot be at once identified with any of those with which we are acquainted. Scientific observers have compared the odour in question to that of phosphorus, nitrous acid, ozone, etc.

There is a case in which the odour was said to be *alliaceous*, or like that of onions, rather than sulphureous; but it has been asserted that sulphur has been actually found on the surface of gilt metal objects blackened by lightning. Some liken it to that of gunpowder. M. Bonjean thinks that sulphide of hydrogen rather than sulphurous acid accompanies the lightning, and assists in blackening the metal. So completely does the idea of the sulphur odour possess the public mind in connection with lightning, that in a case brought before the Academy of Sciences at Paris, of a sailor struck by lightning, on recovering his consciousness he is said to have become sea-sick, and the vomit smelt strongly of sulphur.

We now come to the consideration of the most curious part of the subject—*Lightning Figures*, or the various impressions said to be formed by lightning on striking persons and animals. It has been thought proper to invent a new term for this department of science—if such it may be called—namely, *Keraunography*, or lightning-figures: and M. Andres Poey, Director of the Physico-Meteorological Observatory of Havannah, has exhausted the subject in a very amusing little volume.[1] This curious book gave occasion to an examination of the subject by Mr. C. Tomlinson, Lecturer on Science, King's College, London, before the physical section of the British Association at Manchester in 1861.

---

[1] 'Relation Historique et Théorie des Images Photo-électriques de la Foudre.'

Out of the numberless statements of the kind we select the following :—

In August, 1853, a little girl was standing at a window, before which was a young maple-tree. " A complete image of the tree" was found impressed on her body after a flash of lightning.

A boy climbed a tree to steal a bird's nest; the tree was struck by lightning, and the boy thrown to the ground : " on his breast the image of the tree, with the bird and nest on one of its branches, appeared very plainly."

Hereupon Mr. Tomlinson remarks, that " when boys ascend trees to steal birds' nests, the poor little fluttering parent does not, in this country at least, stop to have its photographic portrait taken." This objection is scarcely valid; certainly on such occasions the fluttering parent does not stop to have her picture taken, but we apprehend that she does sometimes stop to protect her little ones.

Another case is that of an Italian lady of Lugano, who, sitting at the window during a thunderstorm, had the portrait of a flower permanently impressed upon her leg.

During the thunderstorms of June, 1866, similar images of trees, etc., are said to have been impressed by lightning.

By the lightnings which attended the eruption of Vesuvius in 1660, innumerable crosses were imprinted on garments, shirts, and aprons all over the kingdom of Naples.

In 1689, in the church of Saint-Sauveur, at Lagny, in France, the lightning printed some of the words from the cards, as seen on Catholic altars, on the altar-cloth.

At Havannah, the lightning killed a cat which was giving suck to her little ones, and " on her belly was found impressed the figure of a circle two inches and a half in diameter, which was the representation of a larger circle close at hand."

A sailor sitting near a mast was killed by lightning, "and from his neck to his navel there was the image of a horse-shoe, perfectly distinct and of the same size as one which was nailed on the mast."

In numerous cases the images of coins have been impressed—all with greater or less precision ; and, finally, a woman who had charge of a cow was struck down by a flash of lightning which killed the cow, and on her breast was found " a perfectly engraved image of the cow."

In all such statements it is difficult to get the real value of the fact, without which all attempt at explanation must be idle. Such investigations remind us of the joke played off upon certain learned philosophers by Charles II. His Majesty asked them to explain how it happened that if a fish be placed in a tub containing water it would not add to the weight thereof? Most of the philosophers gave learned explanations—at each of which his Majesty shook his head doubtfully. At length a canny Scotch philosopher among the number exclaimed, " But is it a *fact*, Sire? " " No," said the royal wag laughing, " it is *not* a fact."

And so, in the instances before us of lightning-figures, the first objection is that real images of the kind mentioned may not have existed at all,—mere fancy supplying the likeness.

On the other hand, it is quite certain that electrical discharges are capable of producing impressions. For example, a coin is placed on glass, and a stream of sparks from an ordinary electrical machine is poured upon it. About 80 or 100 turns of a two-foot plate machine may be required. On throwing off the coin and breathing on the glass, the image and superscription of the coin are, under favourable circumstances, perfectly reproduced, by the mode in which the breath condenses on the glass. We say, under favourable circumstances, because it is necessary that *a film of matter* such as covers most objects exposed to the air or to contact with the hands, be on the glass; and the action of the electricity seems to be to burn off greater or less portions of this film, coinciding with the greater or less projections of the coin, so that when the breath is projected on the glass, the moisture becomes condensed in a regulated manner according to the regulated action of the electricity on the film; that is, where the electricity has touched the glass and burnt off the organic film, the breath becomes condensed in continuous streams of water, but where the film still remains, the moisture is condensed in minute globules. These images are called *Breath Figures.*

Now, if we discharge a Leyden jar upon a pane of glass, by interposing it between the knob of the jar and that of the discharging rod, we get a breath-figure which may be taken as the portrait of the discharge of a miniature flash of lightning, representing, doubtless, on a small scale, the mode of discharge on the large, where the earth and the clouds take the place of the two coatings, and the air that of the insulating glass. In this experiment, a small Leyden jar and a pane of window glass three or four inches square will suffice. The glass should be held by one corner; and one knob of the discharging-rod being placed on the coating of the jar, the glass in contact with the other knob should be brought quickly up to the knob of the jar, when the discharge will take place—not through the glass, but along its surface, turning over its edge, and passing up the glass to the knob of the discharging-rod on the other side.

Thus, we get two figures, the principal one on the side next the jar, and a second subsidiary figure on the

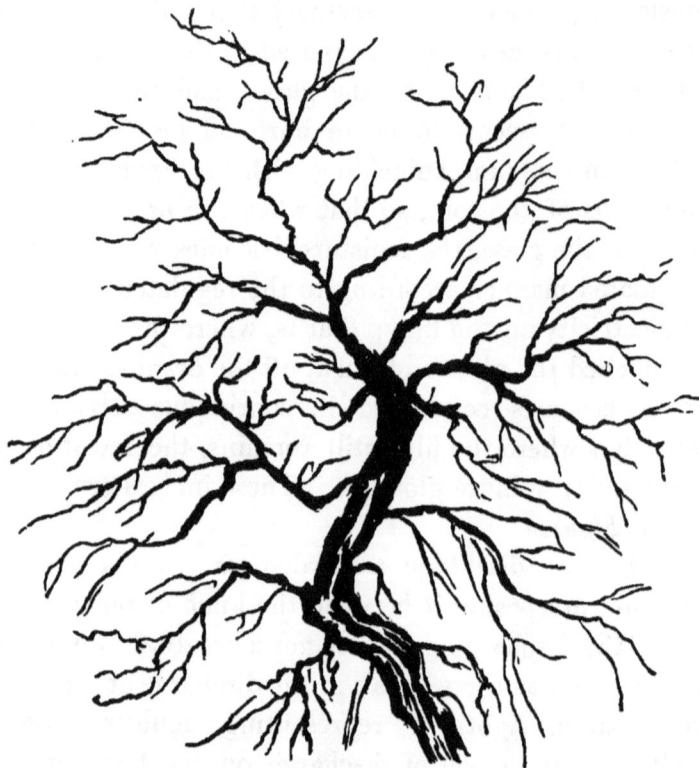

discharging-rod side. These figures come out beautifully by breathing on the glass and holding it up to the light. Wherever the electricity has burnt off the film, the moisture is deposited in unbroken lines; but in the other parts where the film is intact, it is in very minute globules. The following figure, furnished by Mr. Tomlinson, represents roughly a specimen of the principal figure, which is that of a *leafless tree,* or something so provokingly like a tree, that any one who has seen it, and has read the wonderful stories in M. Poey's memoir, exclaims at once, "Here is the origin of the photo-electric figures of lightning!" Mr. Tomlinson showed this experiment to one of the Professors of King's College, and he exclaimed, "There is the branch, and there is the bird's nest, as plainly as possible!"

That a person struck by lightning, while standing under a tree, should have tree-like impressions on his person, would naturally lead an ordinary observer to see "an exact portrait of the tree" in these marks; the blotches are taken for leaves, for a bird's nest, etc., as the case may be. But should the victim be conveyed to a medical man, he would be likely to interpret these ramifications into a case of "ecchymosis," as the doctors call it, and to report accordingly. Doubtless, the lightning itself, or one of its ramifications, prints its own fiery mark on the skin of the victim, and thus produces these tree-like impressions which have excited so much astonishment, and led to so much false description and theory during the last eighty or ninety years. On examining a tree which had been stripped of its bark by a stroke of lightning, Mr. Charles Poolcy, of Weston-super-Mare, found the inner surface of the bark to contain ramified impressions of the lightning, corresponding with those above given. Specimens of this bark were forwarded to Professor Faraday, and are now in the Museum of the Royal Institution of London.[1]

[1] 'On Lightning Figures,' by C. Tomlinson. Also, "Thunder Storm" and "Breath Figures" in the *English Cyclopædia*, by the same author.

So much for lightning-figures, and their explanation, which dispenses with any examination of the scientific objections to the possibility of the truth of the various statements which are universally made respecting such occurrences.

Fulminary tubes, or *fulgurites*, as they are called, have also been objects of popular superstition. They consist of tubes of different lengths and diameters, vitrified within and covered externally with agglutinated grains of sand. Blumenbach was the first to attribute their formation to lightning, which direct observations have since confirmed. Some sailors saw the lightning fall upon sand; they dug and found a tube, blackened with the carbon of the burned vegetables. Bendant, Hachette, and Savart obtained artificial fulgurites by making powerful electric sparks pass into sand mixed with salt, in order to increase its fusibility.[1]

Thunderbolt is a grand poetic and oratorical " flash" phrase, and the idea is inseparable from that of the pagan Jupiter, the Thunderer *par excellence ;* but it is only a metaphor or figure of speech without reality. Lightning is not even a " fluid," as we call it. What it is precisely, is as yet unknown; we only know it by its effects, which are, however, serious enough to warrant the name of " bolt." The term is also applied to meteoric stones or aerolites, before mentioned, and to the small masses of iron pyrites found on the seashore and elsewhere. The " true history," however, of a modern thunderbolt was recently furnished by Mr. Symons, of Camden Road.

On the 2nd of July, 1866, one of the London daily papers startled some of its readers by inserting a letter on the thunderstorm of June 30th, in which the writer

---

[1] Kaemst, *ubi supra.*

said : " At six o'clock it broke out again with increased violence, the rain descending in torrents, and at a few minutes past seven a *thunderbolt* fell in the gutter opposite my house (in Westbourne Park, Notting Hill), and was *smashed to pieces*, one of which I have at the hour at which I write. Numbers of persons are still *searching for the fragments."*

Now, here was the grand mystery explained at last. A live thunderbolt was bagged and ready for exhibition —possibly at the Polytechnic, introduced with the well-known eloquence of Professor Pepper. Of course Mr. Symons, who would go to the world's end if he could to get at the bottom of all that concerns rainfall and the elements, was perfectly electrified by this announcement, and at once abandoning his rain-gauges and rainfall statistics, he started off to hunt up the thunderbolt He called on the writer of the circumstantial letter just quoted, and other residents, found the story amply confirmed, and received several pieces of the thunderbolt (some of which were sold at ten and fifteen shillings each). Nobody had the least doubt of it : besides, it was "in print," and it is astonishing how anything "in print" takes in and satisfies the British public. Mr. Symons, however, was not taken in by the thunderbolt. He made further inquiries, and found that one of the pupils of an analytical chemist had availed himself of his tutor's absence to fill a capsule with materials calculated to burn vigorously and explode in heavy rain, and, during the height of the storm, had thrown the burning mass into the gutter,—so making the thunderbolt in question.[1]

[1] 'Symons's Meteorological Magazine,' June, 1866. This is a very cheap and useful record of rainfall and general meteorology, and its

In connection with lightning must be mentioned certain curious phenomena called *Corposants*. In the 'Edinburgh New Philosophical Journal,' July–October, 1830, is given the following account of one of them by Lieut. A. Milne, of H.M.S. Cadmus, in latitude 34° 40′ S., longitude 54° 50′, in September, 1827. "About ten o'clock at night, while the lightning continued to rage, and to extend itself around the horizon, I observed a light at the extremity of the vane-staff at the mast-head; and shortly after, another on the weather-side of the foretopsail yard. One of the midshipmen went aloft to discover its position. He found it attached to an iron bolt on the yard-arm, its size rather exceeding that of a walnut, having a faint yellow cast in the centre, and approaching to blue on the exterior edge. He applied his hand to it, on which it burnt with a hissing noise, resembling the burning of a portfire, at the same time emitting a dense smoke without any sensible smell. When he applied the sleeve of his wet jacket it ran up it, and immediately went out,—the electricity being conducted another way by the water. The light on the vane-staff retained its position for upwards of an hour." In the month of December, the same year, a second appearance of the Corposant, when off the coast of Patagonia, was observed, attached as well to a spindle of hard wood as to one of copper or iron. "After the above phenomena," says the lieutenant, " we had always very bad weather, commencing with heavy and sudden squalls, generally from S.W., but varying a few points each way, and settling in a few hours to a steady gale,"—that is, after each display of electric power accumulated in the

editor, Mr. Symons, of 136, Camden Road, London, will be thankful for all information on the subject.

atmosphere of these latitudes, there was a *change of wind and weather*. These strong S.W. winds come from the land, and are called *Pamperos*.

This phenomenon appears to be the same as that observed by the ancients and called *Helena*, corrupted by the modern sailors into *St. Elme's* Fire.[1]

On the 20th February, 1817, a surgeon at Leadhills, in Scotland, during a snow shower, found his horse's ears suddenly made luminous and the brim of his hat appearing as if on fire.

---

[1] The name *Corposant* is the Italian *Corpo Santo*, "holy body." The Italian sailors are fond of giving pious names to ominous objects. Thus, they named the bird Stormy *Petrel* from *Pedrillo*, "little Peter"—from St. Peter walking on the sea; and doubtless its other name, "Mother Cary's Chicken," is a corruption of one of the names of the Virgin Mary—*Mater cara* or *carissima*.

CHAPTER XXI.

ARTIFICIAL RAIN; SHOWERS OF BLOOD, CORN, SULPHUR, FROGS, BURNT PAPER, SOOT OR BLACK RAIN, ETC.; HONEY DEW.

THERE are not a few facts on record which seem to show that in seasons of great drought we might be able to bring down fertilizing showers in abundance. We will begin with the facts, and then venture on our suggestions.

After a succession of very hot days, when the air is calm and highly charged with aqueous vapour, any violent concussion will force a part of it into a higher colder medium, where it will be rapidly condensed into rain, which falling through the under clouds, will cause a further deposition, that will be propagated, and produce a sudden general shower, frequently accompanied with lightning and thunder.

Thus, the firing of guns will, under such circumstances, produce a change of weather from drought to stormy, of which the following instances may be sufficient to give a satisfactory demonstration.

During the siege of Valenciennes by the Allied Army

in July, 1793, the weather, which had been remarkably hot and dry, became violently rainy after the cannonading commenced.   Two hundred pieces of heavy ordnance were employed in the attack, and one hundred in the defence of the city, the whole of which were frequently in action at the same time.

At the Battle of Dresden, August 27th, 1813, the weather, which for some days previously had been serene and intensely hot, now suddenly changed, vast clouds filled the skies, and soon the surcharged moisture poured itself out in a torrent of rain.

Lord Exmouth, in a letter to his brother, stated that the night following the bombardment of Algiers, in August, 1816, there was a dreadful storm of thunder, lightning, and rain, as heavy as he ever witnessed.

Lastly, at Waterloo, as Siborne relates, the weather during the morning of June 17th, 1815, had become oppressively hot.   It was now a dead calm; not a leaf was stirring; and the atmosphere was close to an intolerable degree, while a dark, heavy, dense cloud impended over the combatants.   The 18th Hussars were fully prepared, and awaited but the command to charge, when the brigade guns on the right commenced firing, for the purpose of previously disturbing and breaking the order of the enemy's advance.   The concussion seemed instantly to rebound through the still atmosphere, and communicate, like an electric spark, with the heavily charged mass above.   A most awfully loud thunderclap burst forth, immediately succeeded by a rain which has never probably been exceeded in violence even within the tropics.   In a very few minutes the ground became perfectly saturated.

After the middle of August, none of the inhabitants

are suffered to ascend to the tops of the surrounding
Himalaya mountains, or to use firearms in the neighbourhood of the villages; as the occurrence of either of
these events, at that time, is found from experience,
generally, to occasion a fall of snow above, and a frost
below; by the latter of which, the ripening crops would
be wholly destroyed.

The inhabitants of the Valley of Chamouny, in Switzerland, are aware that the discharge of a fowling-piece,
or even loud speaking, would bring down an avalanche,
or break some of those huge icy pinnacles known by the
name of *aiguilles,* or needles, rising out of the glacier.
The firing of a musket by the discovery ships in one of
the bays of Spitzbergen, shivered in pieces an enormous
iceberg, whose fragments covered a *square mile* of the
surface of the sea. With regard to the fall of snow and
the frost, it must be recollected that any sudden concussion of the air, when in a calm state, at or below the
freezing-point, will produce an instantaneous congelation
of the suspended vapour—just as water cooled below
the temperature of melting snow will remain liquid, but
is immediately congealed on the slightest agitation,—or,
as Glauber's salt dissolved in warm water will, if shaken
when cold, crystallize at once, and assume a solid form.

Natural phenomena exactly in point may be quoted.
The disturbance of the air arising from earthquakes,
when the atmosphere is highly charged with aqueous
vapour, is generally followed by rain. After a great
drought and heat, a volcanic eruption will change the
weather into rainy and stormy; possibly from the sudden
expansion of the air upwards into a colder region, where
its aqueous vapour is condensed into rain, which, falling upon the hot ground, cools it by evaporation, and

produces a lower temperature. This occasions stormy weather by the commotion of the atmosphere arising from the mixture of airs of different densities, which, owing to their moisture, produce clouds that render the air above them extremely cold,—for a cloud is pretty much like an iceberg in the skies.

The violent eruption of Vesuvius in the year 79, which buried the cities of Pompeii and Herculaneum under a shower of lava, pumice-stones, sand, and ashes, was accompanied with heavy rains. The same occurred in the eruption of March 28th, 1828. During an eruption of a volcano in Iceland in 1793, not only did rain fall in torrents, but also hail in showers. The terrible eruption of Vesuvius in October, 1822, was preceded by a drought which spread desolation in the fields; but towards the end of the phenomenon the volcanic thunderstorm caused an extremely heavy and long-continued rain. In all countries the cessation of an eruption is characterized by a similar result. Humboldt states that when a volcano bursts out in South America during a dry season, it sometimes changes it to a rainy one.

From all these facts we venture to suggest the possibility of mitigating the effects of drought. It is a well-known fact that, in very hot calm weather, the burning of woods, long grass, and other combustible materials, produce rain. The instances given of it, particularly in North America, are too numerous for insertion. Very extensive fires in Nova Scotia are so generally followed by heavy floods of rain, that there is some reason for believing that the enormous pillars of smoke have some share in producing them.

It is requisite, in order for this circumstance to take

place, that the dew-point should be sufficiently high before the fire; in which case, the aqueous vapour will be so much expanded as to rise into air below the temperature of that point, where it will condense into clouds, which will render the air above them so cold as to cause it to descend, and by that means, increase their quantity. This process will continually produce an addition to the moisture, which will ultimately become so copious that, by the conglomeration of its particles, their weight will cause them to descend in rain-drops.

There may be few or no localities in our agricultural districts exactly adapted to produce rain by a conflagration of waste materials, long grass, etc.; but we see no reason why the Government might not permit the trial of artillery in various parts of the country during seasons of protracted drought—when the skies are often crowded with tantalizing clouds which will not send down their moisture. This would be a much better use of gunpowder than when blazed away on field-days and in bloody battles. A hundred or two of our guns might be sent off and posted in various parts of the country, and fired after the manner of a siege, when no doubt the same result will ensue as on the occasions we have instanced,—at the same time giving " practice" to the men at least in working the guns. At any rate, the main facts are worth knowing, as they tend to explain the phenomenon of rain.

In 1806, Williams proposed to electrify the whole atmosphere of Great Britain one mile in height, by ordinary electrifying machines, so as to render the seasons more propitious to the health of our growing crops, and to counteract unusual drought. He believed that, by keeping the air strongly electrified in the

spring season, we might perhaps be able to prevent the
deleterious effects to our fruits and vegetables, arising
from ungenial weather in April and May; and proposed
for this purpose the erection of one or two buildings
furnished with the requisite apparatus in each county, on
a heath or in a situation devoid of trees and buildings.
He reasoned as follows :—" Two revolutions of an excited
electrical machine—which may be performed in two
seconds of time—will electrify the air of a room 24 feet
long by 18 feet wide and 13 high, as strongly as I
ever found fog electrified in the month of September,—
provided a lamp or candle be placed on the insulated
conductor, so as to diffuse the electric matter; and this
electricity is not wholly re-absorbed by the walls, floor,
or ceiling of the room, in less time afterwards than
one quarter of an hour. Suppose, therefore, a building
erected and furnished with machinery, somewhat similar
to a cotton or silk-mill, and that the various movements
consisted of cylinders or plates of glass, fitted up with
rubbers, etc., for exciting electricity; and so arranged
as to convey the electric matter into an insulated upright bar, terminating without the roof of the building, in a large lamp, or a series of lamps and points
for again diffusing the electrical matter in the circum-
ambient air. 1 find, by calculation, that a force adequate to work a common pair of millstones would give
motion to twelve hundred such electrical cylinders or
plates of glass. If, therefore, one cylinder, in two seconds of time, will electrify so many cubic feet of air
contained in a room 24 feet by 18 and 13 feet high, it
might be easy to calculate the quantity of vapour needed
for any given space and height, expanse being also attended to, in any given time,—the number and power of

such apparatus being previously ascertained. A calculation might thus be made to decide what number of
machines would be adequate to electrify the whole
atmosphere of Great Britain one mile in height; for it
does not appear that dense vapours ascend much higher
than this in our climate; and the dry state of the transparent air would preserve the insulation; so that the
electricity thus given to the atmosphere would not diffuse
its influence far above the vaporous regions."

With respect to the showers of blood, corn, sulphur,
etc., on record, we must begin with quoting the answer
of one of the most distinguished naturalists of the age
to some one who assured him that he had seen one of
these phenomena with his own eyes:—" It is fortu-
nate," he replied, " that you have seen it, for now I
believe it: had I seen it myself, I should not have believed it."

In the Middle Ages, when " the anger of Heaven"
was almost the only means of bridling or frightening
men, " showers of blood" were frequently reported.
The rain so designated has fallen in modern times, and
the microscope has explained the mystery, tracing the
red colour to innumerable microscopic vegetable " cells"
or animalcules, or to a red powder containing inorganic
matter coloured with iron (which gives the colour of
blood) or hydrochlorate of cobalt. Red snow is found
on the Alps and the Pyrenees, and in Baffin's Bay; but
the microscope proves the colour to result from red
granules the nature of which identifies it with a plant—
the *Hæmatococcus nivalis*. Showers of corn have been
reported, but scientific investigation has proved the
supposed corn-seeds to be the tubercles of the *Ranunculus .Ficaria*. In the middle of June, the leaves and

stalks of this plant dry up, and nothing but roots remain, consisting of small tubercles fixed to feeble radicles. A heavy rain draws up these tubercles, separates them from the root, and carries them to inclined places, and so they are frequently found after rain,—but no one has yet seen them fall with it. As gusts of wind may transport seeds and even fruit, they may be observed after tempestuous rains, and such have frequently been seen, but always referable to the cause mentioned. The " showers of sulphur" consist of the pollen of certain flowers and of pines in particular, swept off by the wind, and precipitated with the rain. Frogs, fish, snails, etc., appear to fall with rain, but the only established fact is that they are found in numbers in fields after rain. If not wafted by the winds they are merely drifted by the rains, or attracted out of their retreat by the moisture.[1] Even a pike has been known to seek another portion of his element by wriggling over the grass (moistened by dew) ; and there are fishes which actually work their way out of drying ponds into " more water," nay, one species actually climbs a tree, apparently to get a glimpse of the direction he is to take in order to find " more water" ! Nature has given him a hook on his snout and another near his tail, and the instinct to use them in this exploring escalade. Of showers of soot or black rain we have recently heard much, and a volume has been written on the subject,[2] but with Mr. Symons (our indefatigable authority on rainfall) we believe their origin is the chimneys, to which everybody seems inclined to refer them. The same expounder announces

---

[1] Kaemst, *ubi suprà*.

[2] 'The Scottish Black Rain Showers', etc., by the Rev. James Rust. W. Blackwood and Sons.

the fall of "pumice-stone shoals" in Great Britain; but again, with Mr. Symons, without denying that rain may bring volcanic dust (instances of which are on record), we expect to hear of the phenomenon not from one parish or even one county only, but from observers in all parts of the kingdom; therefore, for the credit of the nineteenth century, any man asserting such facts should rigorously sift them ere he announces them to the public.[1]

No doubt whirlwinds will carry up everything within their reach and transport them to a distance; but whatever falls in a shower must be of the earth,—unless it be " meteoric dust," or matter resulting from the destruction of meteors, falling stars, etc.

The following is a curious instance of the performance of a whirlwind on the 30th of June, 1866, in the neighbourhood of Tadcaster:—" It appeared like a flexible column of smoke standing out in bold relief against a large thundercloud in the east. In its course it unroofed a farmhouse, carried large haycocks, stones, and limbs of trees into the air, swept up tons of water from the river Wharfe, and gave other evidences of its power. Its track was about 50 yards wide, and made a noise resembling that of a railway train in rapid motion. Another correspondent says on Saturday, about noon, at Ozendyke, near Ulleskelf, a singular-looking cloud, of the shape of an elephant's trunk, appeared in the air just overhead. It was wide at the top, and tapering off to a very small point at the bottom, within 150 yards from the earth. This appearance continued about 15

---

[1] See 'Symons's Monthly Meteorological Magazine,' June, 1866, for a refutation of the statement.

minutes, all the time revolving very rapidly, and making a noise like a large manufactory at work. It was apparently extending and getting nearer to the earth with its tail, when all at once it seemed to burst and emit what we thought water, causing a great mist, though it was more like a large steampipe when the steam is let off. Immediately after there was an alarming whirlwind, or tornado, tearing up everything before it. The first thing we saw it strike was a large poplar-tree in Mr. Farrar's farm, which it nearly stripped of its branches, breaking off the boughs, many of which would weigh more than a quarter of a ton each. These it carried more than 100 yards high, and afterwards dropped at a distance of 100 to 300 yards, forming a circuit round the tree. It afterwards went in the direction of Mr. Farrar's orchard, where it did great destruction to the fruit-trees. These it twisted round like a corkscrew, although many of them are as thick as a man's body. It afterwards passed in the direction of the river Wharfe, which it crossed, taking a large quantity of water up to an immense height. After crossing the river it laid the meadow in Bolton Ings (about 100 yards wide) quite flat in a straight direction. It then recrossed the Wharfe near Bolton Clough, and was seen again to take a quantity of water up higher than the loftiest tree in the neighbourhood."

Luke Howard witnessed "a continuous shower of fragments of *burnt paper*, descending from an elevation which the eye could not appreciate;" but it originated from the burning of the Custom House (12 Feb. 1814), distant in a right line about five miles S., wafted by a S.E. breeze.

Analogous to the popular notions, we have instanced,

is the supposed origin of *Honey Dew,* namely, that during the night, a shower of mellifluous fluid falls on
the leaves of certain plants. Obviously, if this were
the case, the gravel-walks of gardens and other shrubs
and plants contiguous to those affected, would be covered
also; it happens that only particular plants, not adapted
to resist sudden variations of temperature, are subject
to this disease, for such it is to all intents and purposes.
In the months of April, May, and June, if we get a few
hours or days of bright sunshine, the thermometer rises
as high as in the South of France, or even Italy. This
heat is frequently succeeded, with us, by frosty or very
cold nights, and perhaps the next day or two is followed
by hailstorms, or cold, cloudy weather. Certain plants
cannot stand this sudden change. The living principle
or excitability of the plant is in the first instance thrown
into great action, during the continuance of heat; and
this being suddenly withdrawn, a torpor succeeds; the
circulation of the juices in the finer vessels of the leaves
and tender shoots is obstructed, while the absorbent
vessels of the roots, by the assistance of the contractile
power of the outward ring of woody fibre not having received injury, continue to propel the rising sap, which,
arriving at the injured leaves and young terminal shoots,
the proper secretions and excretions are here intercepted,
new materials and new vessels are produced. Similar
consequences ensue to animal life by what is called
"frost bite,"—namely, a local and temporary torpor;
but, as the heart, like the trunk and root of trees, is not
injured, this organ propels the blood to the injured part,
where, meeting with obstructions, " inflammation," with
the consequent production of new vessels, and new secretions, succeeds. Thus, the saccharine, gummy substance

called honey-dew, observable on the foliage of plants in spring and early summer, is the product of disease caused by cold or frost. It is the diseased state of the vegetable absorbent vessels, which thus deposit the juice in question, termed honey from its taste, and dew from its appearance. Such being the case, all external applications are useless. If anything could effect a cure, it must act by " lowering the inflammatory state," but no medicine has as yet been given to plants. The timely blanket of matting thrown over them during the nights of such hot days would prevent the evil.

CHAPTER XXII.

A METEOROLOGICAL GARDEN AND FLORA'S CLOCK.

It was reserved to the great Humboldt, who united in himself the sum of human knowledge, to show that meteorology and botany, so distant in the hierarchy of the sciences, are sisters in the harmonious unity of nature. In his numerous travels he had everywhere seen vegetation modified or changed when the climatological conditions were not the same; he studied the relations existing between the physiognomy of American flowers and the climates to which they correspond, and thus created *Botanical Geography*. It is not, however, merely in their general relation to meteorology that plants are interesting, but also in the particular function which many of them possess enabling them to foreshow changes of the weather.

The celebrated botanist, Linnæus, is said to have possessed such a knowledge of the periods and indications of flowers, that he required neither a watch, a calendar, nor a weather-glass. He has enumerated forty-six flowers which possess this kind of sensibility, and divides them into three classes.

1. *Meteoric Flowers*, which less accurately observe

the hour of folding, but are expanded sooner or later, according to the cloudiness, moisture, or pressure of the atmosphere.

2. *Tropical Flowers*, which open in the morning and close before evening every day ; but the hour of their expanding becomes earlier or later as the length of the day increases or diminishes.

3. *Equinoctial Flowers*, which open at a certain and exact hour of the day, and for the most part close at another determinate hour.

That temperature, moisture, and atmospheric pressure have everything to do with the growth of plants is incontestably proved, not only by experience but their very organization. Plants usually expand their flowers well and perfectly on fine days, that is, in the good conditions of temperature, pressure, and adequate moisture in the soil. Every practical gardener and farmer knows the invigorating influence on his plants and crops, of the first hours of the spring or summer day, when the sun first shoots his horizontal beams above the horizon, dispelling the vapours of the night with his genial rays. This has been attributed to a mysterious substance to which the name of *ozone* has been given, concerning which there has been immense discussion, but which seems to be nothing more than oxygen in a condensed form, or, as some say, electrified. We believe we have reason to say that little or nothing is really known either of its origin or mode of action. Certain it is, however, that there is a peculiar chemical action set up by those early beams of sunshine in passing through the atmosphere, calling into vigorous activity certain conditions of oxygen, not always existing in such force and quantity, creating what, for want of a better name, is called ozone.

Not only to vegetable life is the peculiar effect in question useful, but also to animal life, which equally feels its beneficial influence. Any one who is in the habit of being out in the open air shortly after sunrise, cannot fail to have noticed the exhilarating effect produced upon the nervous system, and the pleasurable sensations which seem to flood the early riser as he inhales the perfume of the "incense-breathing morn." The same phenomena are felt after a thunderstorm has "cleared the air," as we say, but of what we know not, excepting generally all that does not constitute the proper qualities of breathing air. And so it has long been proverbial that early rising and walking in the open air are conducive to the beauty of the complexion ; for, the breathing of pure air is as essential as the best of food, and enables the latter to establish that excellent state of health of which the complexion is a probable, though not an infallible criterion.

Doubtless the action called electricity is the prime mover in all vitality, whether vegetable or animal. Its influence throughout creation is twofold : indirectly it causes the precipitation of aqueous vapour, as in thunderstorms ; it is connected with the distribution of heat, and the variations of atmospheric pressure ; directly, its influence, both on the entire animal and vegetable creation, is not less powerful and important in stimulating the nerves of animals and promoting the circulation of the organic juices of plants. Whilst the indirect influence is only occasional, varying in degree in different parts of the world, and in some seasons is much greater than others, its direct influence is ever at work, pervading every particle of matter.

Electricity is of two kinds, to which the terms negative and positive have been applied, the positive kind being that usually present. During a storm both kinds are usually exhibited, and generally change from the one kind to the other with great rapidity. In fine and calm weather the electricity is invariably of the positive kind, and the tension is the strongest in the evening. During fog the tension is usually very strong, and sometimes changes from positive to negative. It is a pretty sure sign of a run of bad weather when the electroscope indicates the presence of the *negative* kind. During falls of snow the kinds alternate repeatedly; during hail, permanent negative electricity is shown. The amount of electricity, like all other meteorological phenomena, has its periods of daily *maxima* and *minima*, as well as longer periods as defined from season to season. The diurnal range is as follows:—At sunrise, the atmospheric electricity is weak; it continues to increase as the sun rises higher above the horizon, and the vapours are collected in the lower strata of the atmosphere. This increase of the tension, with the age of the day, goes on, in summer till six or seven o'clock, in the spring and autumn till about eight or nine, and in the winter months till ten or twelve in the day. After having reached its maximum tension, it generally decreases very rapidly, and by about two o'clock in the afternoon it is usually not much stronger than at sunrise. The minimum occurs, in summer, between five and six in the evening, and in winter at about five o'clock. The minimum tension is of longer duration than the maximum. As the sun declines in the heavens, the electricity begins again to increase, especially at the moment of the sun's reaching the horizon, goes on increasing during the close of the evening, and arrives at its second maximum some two or three hours after sunset. At this period the lower strata of the atmosphere are loaded with vapours, as in the early morning, and night dews fall. From this time 'the tensions get weaker and weaker until the next morning. During fine weather the positive electricity is stronger in the winter than in summer, and varies regularly between those two periods.

To study the electricity of the clouds Franklin was the first to employ the electric kite. He fastened a common kite to a ball of packthread, either wetted, or containing a fine metallic wire; when the kite rose, he placed the ball of thread in connection with an electrometer. This experiment is very dangerous; and modern philosophers have recourse to the electroscope for the purpose of ascertaining atmospheric electricity.

This is the simplest instrument for ascertaining the electric condition of the atmosphere. It consists of two equal pieces of gold leaf suspended from a brass support, and enclosed in by a glass covering, which not only protects it from the air but also "insulates" or preserves it in a neutral condition. The brass-support is surmounted by a metallic rod, not less than two or three feet in length, with a clip at the top to hold a lighted piece of cigar-fusee or touch-paper.

The action of this instrument is very curious. The electricity of the air is collected by the substance undergoing combustion or burning, and conducted by the rod to the gold leaf, when the pieces, being similarly electrified, separate more or less according to the amount of electricity present. The instrument is founded on the law that bodies similarly electrified repel each other, but when dissimilarly electrified, they attract each other.

To make an observation, we place the instrument in

the open air and light the piece of cigar fusec, in order
to collect the electricity whose kind we wish to discover.
For this purpose we provide ourselves with a rod of glass
or a stick of sealing-wax. A rod of glass, when rubbed

briskly with a silk handkerchief or piece of woollen
cloth, becomes *positively* electrified, or " excited," as it is
termed. On the other hand, a stick of sealing wax,
treated in the same way, acquires *negative* electricity.
Consequently, the kind of electricity in the air may be
ascertained by the effect of either the glass-rod or the
stick of sealing-wax upon the electrified or excited gold-
leaf. If, therefore, the excited glass-rod be presented to

the cap of the instrument, and it causes the pieces of gold-leaf to diverge still further, the electric state of the air must be similar to that of the glass, that is, *positive;* if they approach, it is *negative.* On the contrary, if a stick of sealing-wax be used, the pieces will be repelled more apart if they have acquired negative electricity from the air; and they will converge if they have a positive charge. Nothing is more curious than to witness the motion of the gold-leaf under the mysterious influence of causes apparently incapable of the result. By means of this very simple instrument, we can readily ascertain the electric condition of the lower air at any time. The intensity of the electricity of the air is affected by the season of the year, and by the prevalent character and direction of the winds; it varies also with the elevation of the strata of air, being in general greater in the higher than in the lower regions of the atmosphere. The intensity is generally greater in winter, especially in frosty weather, than in summer, and when the air is calm than when winds prevail. North winds give positive, and south winds negative effects. Whether the processes of vegetation and vegetable evaporation have any influence upon the atmospheric charge is quite undecided; but that trees and other plants draw off the electricity of the air, and are perpetually tending to neutralize it with that of the earth, is a well-established fact. Whether their own juices and secretions undergo any change during the process is, however, a point for further inquiry. The theory which ascribes the development of atmospheric electricity to *evaporation* receives considerable support from the fact that the eruptions of volcanoes, in which enormous volumes of steam are discharged into the air and condensed, are constantly accompanied by discharge of thunder and lightning, and the most violent storms occur at times, and in situations where the largest quantities of vapour are generated.[1]

From all these causes—temperature, moisture, atmospheric pressure, and electricity—it is not to be wondered at that plants give signs of approaching changes of weather. Indeed, some naturalists not only attribute to the sensitive-plants (the *Mimosa* genus) a nervous principle which receives impressions externally, but a similar endowment more or less to all plants in general. Sir J. E. Smith, the botanist, would not pluck a flower for fear of giving pain.

Others object to smelling flowers, owing to multitudes of animalcules which flourish in every flower. " The fragrance of the carnation," says Sir John Hill, " led me to enjoy it frequently and near ; and while the sense of smelling was satiated with the powerful scent, the ear was constantly entertained by an extremely soft and agreeable murmuring sound. I distended the lower part of the flower, and placing it in full light, could discover, by means of the microscope, troops of insects frisking and capering with wild jollity among the narrow pedestals that supported its leaves. I admired their elegant limbs, their velvet shoulders, their backs, vying with the empyrean blue, and their darting eyes of fire."

All plants usually expand their flowers well and perfectly on fine days, but many sorts close their petals against the coming of rain ; hence we may often judge of the impending weather early in the morning by noticing the flowers. All plants are very apt to flag and droop before rain, particularly in summer, when after long dry weather, the wind which is to bring up the rain begins

[1] Daniell, Elem. of Meteorol.

to blow. Their means of communication with the air consists of an infinite number of tracheæ or air-vessels, which are visible in the leaf of the scabious, vine, etc. Hence it is, that all wood, even the hardest and most solid, swells in moist weather, as before explained.

All flowers have different temperatures, all higher than that of the atmosphere; and plants being the first decomposers of minerals into their elements, and also assimilating the constituents of air and water, must be considered Nature's chemists, ever at work by means of the universal forces above enumerated. As, therefore, all changes in the weather depend upon the alternations of these forces, and as plants are affected by them, it is evident that any garden may be turned to the purpose of meteorology and weather-wisdom by the daily observation of its vegetable hygrometers, thermometers, electrometers, and barometers.[1]

There are five plants which have been observed from time immemorial for the signs of the weather—the Dandelion, the Trefoil, the Pimpernel, Chickweed, and the Siberian Sowthistle.

---

[1] According to Professor Lindley, not one gardener in a hundred formerly remembered that the atmosphere contained any water at all; not one in a thousand was aware that the quantity floating in it could be actually measured and read-off, like temperature, by a thermometer scale; not one in ten thousand considered that the healthiness of a plant depended essentially upon the relative amounts of atmospheric moisture, light, and temperature, and the peculiar condition of the plant exposed to them. If the plants were flagging, the roots were watered; if some very particular cause appeared to call for it they were syringed, and that was all. When the practice of steaming plants was first resorted to surprise was felt at the results obtained, and the opinion was that it was the *warmth* of the vapour that produced so beneficial an effect. It was Daniell who taught the gardening world that the amount of moisture permanently surrounding plants is of the very first importance.

1. The Dandelion is a very common plant, which flowers early, and remains in bloom more or less all the year. The general flowering, however, takes place about the 8th of April, and for a month it bespangles the fields, mixing agreeably with the Daisy. The down of the dandelion closes for bad weather, but expands for the return of sunshine; the down of other plants may be observed for the same indications.

2. The Trefoil, according to the observation of the great Lord Bacon, grows more upright, with a swelling stalk, against rainy weather; and the same may be said of the stalks of most other plants, though not so conspicuously as in the trefoil. Before showers the trefoil contracts its leaves, as does the Convolvulus and many other plants.

3. The Pimpernel is the *Anagallis arvensis* of Linnæus, and is found in our stubble-fields, and in gardens, flowering in June, and continuing all the summer. When this plant is seen in the morning with its little red flowers widely extended, we may generally expect a fine day; on the contrary, when the petals are closed, rain will soon follow. This is the plant which Lord Bacon seems to refer to under the name of *Windcope*, and which has also been styled the *Poor Man's Weather-glass*.

4. Chickweed is said to be an excellent weather-guide. When the flower expands freely, no rain will fall for many hours; if it so continue open, no rain for a long time need be feared. In showery days the flower appears half concealed, and this state may be regarded as indicative of showery weather; when it is entirely shut we may expect a rainy day.

5. If the flowers of the Siberian Sowthistle remain open all night, we may expect rain next day.

2 A

We have no doubt that if the subject were systematically studied in daily observation, almost every plant would be found to indicate more or less conspicuously all coming changes of the weather, and so it is obvious that a new charm or interest might be given to our gardens, an examination or passing inspection of which in the morning " before leaving for town," would lead us to infer whether "we had better take an umbrella or not "—a matter of frequent doubt in our changeable climate.

Besides being natural meteorological instruments, plants have also shown themselves capable of serving as clocks or watches. The term *Flora's Clock* has been applied to this curious phenomenon, denoting the periodical opening of flowers, whereby the hours of the day are indicated.

As the opening of flowers depends upon temperature, the main fact need not surprise us, but it is very curious to discover that the degree of temperature required in every case is so accurately measured by nature as to take place at fixed periods of the twenty-four hours. Accordingly we find that the flowers of one country do not.open at the same hour in others. Thus, an African plant which opens at 6 o'clock, if removed to France, will not open till 9, nor in Sweden till 10 ; that is, until it gets the requisite temperature. Those which do not open in Africa till noon, do not open at all in Europe in the open air, because, of course, their requisite temperature is never attained.[1]

---

[1] In connection with this fact, we may state that by watering all plants with tepid water, their bloom will be vastly accelerated, and we may secure a magnificent blow in our gardens even during untoward seasons, whilst our neighbours are waiting for the sunshine. Of course all greenhouse plants should have tepid water. Cold watering is always apt to cause the rootlets to shrink and check vegetation. Man prefers cold water, but evidently the plants do not.

1. The Goatsbeard opens with sunrise and closes at 10.

2. The Garden Lettuce opens at 7, and shuts at 10.

3. The Yellow Star of Jerusalem (*Tragopogon pratensis*) and the Purple Star of Jerusalem (*T. porrifolius*) close their flowers exactly at noon.

4. The Mouse-ear closes at half-past 2.

5. The Cat's-ear closes at 3.

6. The Prince's-leaf, or Four-o'clock, opens at 4.

7. The Evening Primrose (*Œnothera biennis*) opens at sunset and closes at daybreak; it opens with a snapping noise.

A species of Serpentine Aloes, without prickles, whose large and beautiful flowers exhale a strong odour of the vanilla during the time of its expansion, is or was cultivated in the Imperial Gardens at Paris. It does not blow till towards the month of July, and about 5 o'clock in the evening, at which time it gradually opens its petals, expands them, droops, and dies. By 10 o'clock the same night it is totally withered—to the great astonishment of the spectators.

The Cerea, a native of Jamaica and Vera Cruz, expands a beautiful coral flower, emitting a fragrant odour for a few hours in the night, and then closes to open no more. The flower is nearly a foot in diameter, the inside of the calyx of a splendid yellow, and the petals of a pure white. It begins to open about 7 or 8 o'clock in the evening, and closes before sunrise in the morning.

The Dandelion must be mentioned again : it opens, in summer, at half-past 5 in the morning, and collects its' petals towards the centre about 9 o'clock. It also possesses the very peculiar means of sheltering itself from the extreme heat of the sun, as it closes whenever the heat becomes excessive.

Such are a few of the most remarkable horary flowers; but no doubt most flowers are more or less " particular" in their hours of opening, depending as the fact does on the requisite temperature, and therefore such observations may be turned to meteorological account in connection with the daily temperature and the seasons.

Other conditions, however, besides temperature, are required for the time of blooming; and of all the propensities of plants none seem at first more unaccountable than the different seasons in which their blossoms appear. Some produce their flowers in winter, as the Christmas rose; others in February, as the elegant snowdrop; in March, the crocus; in April, the sweet-scented violet, peeping through the thorn; in May, the cowslip perfumes our meads; and June is crowned with all the varieties of the unrivalled rose. Thus, through the varying year, till after most plants have formed their seeds, fade, and decay, appears the beautiful *leafless* flower—the winter-crocus. This common circumstance is among the thousand wonders of creation, and it would perhaps be as difficult satisfactorily to explain it as the most rare or stupendous phenomena of nature.

The *awns* of barley, wheat, rye, and such plants, are meant as beneficial auxiliaries or conductors of electric matter for the maturation of the seed; and hairs probably perform a similar kind office on the leaves of hirsute or hairy vegetables. If we deprive some plants of hairs and others of their awns, we should probably affect the growth of the one, or the perfecting of the seed in the other. If we immerse a dry leaf in water just drawn from a spring or pump, which contains a quantity of air, innumerable globules constantly appear on every point. The extremities of these points attract the particles of water less forcibly than those particles attract each other; hence, the contained air, whose electricity was but just balanced by the attractive power of the surrounding particles of water to each other, finds, at the point of each hair or fibre, the resistance to its expansion diminished; it consequently expands, and becomes a *bubble*. The rays of the sun, being partly refracted and partly reflected by the surfaces of these minute bubbles, must impart to them more heat than the transparent water, and thus facilitate their ascent by a further expansion. That the points of vegetables attract particles of water less than they attract each other, is evident from the spherical form of the dew-drop frequently visible in the evening on the points of grass.

In concluding this pleasant topic of the flowers, we cannot help remarking on the barbarous names given to them by our modern botanists. We need not be reminded of the adage about the rose smelling as sweet as ever, call it by what name you please; but still we do not see why ugly, unpronounceable names should be given to sweet pretty things. The old monks, who were our first botanists in England, were much happier in their floral nomenclature in giving some plain significance to the names applied to our gems of the garden. With them the *Tussilago fragrans* was the *Shepherd of Madonna*, in commemoration of the shepherds who awaited the delivery of the Blessed Virgin, as Milton represents them, in a rustic row, or of the Magi, the wise men of the East. The *Galanthus nivalis* (Snowdrop) was *Our Lady of February*, referring to Candlemas Day, or the Purification of the Virgin. At the Reformation the name was corrupted into *Fair Maids of February*. The *Cardamine pratensis* was the Ladysmock, or *Chemise de Notre Dame*, regularly flowering on old Lady Day, April 6th. The *Narcissus Pseudo-Narcissus* (Daffodil), from blooming all Lent, was called the Lent Lily. The *Galium cruciatum* was the Cross-flower, blooming about Holy Cross Day, May 3rd. The *Clematis Vitalba* was the Virgin's-bower, or Traveller's-joy, blooming on the festival of the Virgin's Assumption.[1] The Sunflower was St. Bartholomew's-star, blooming on the saint's day. The *Bellis perennis plena* was the Herb Margaret, blowing about Feb. 22, St. Margaret's Day. *Gratiola* (Hedge Hyssop) was St. John's-wort, subsequently changed into *Hypericum*. The Iris was *Fleur de St. Louis; Ricinus* was *Palma Christi; Viola tricolor* was Herb Trinity ; *Sweet William* was Herb St. William ; *Trichomanes* was *Our Lady's Hair; Flos Jovis* was God's Flower. There were also Job's Tears, Our Lady's Laces, Our Lady's Mantle, Our Lady's Slipper, and of course Monk's Hood, Friar's Cowl, and St. Peter's Herb, etc. In short, as observed by an enthusiast on the subject: "Go into any garden, I say, and these names will remind every one at once of the knowledge of plants possessed by the monks, most of them having been named after the festivals of the saints' days on which their natural time of blooming happened to occur; and others were so called from the tendency of the minds of the religious orders of those days to convert everything into a memento of sacred history and the Catholic religion which they embraced."[2]

[1] We find the name *Flammula Jovis* also applied to *Virgin's-bower.*
[2] 'The Catholic Friend,' *apud* Forster.

CHAPTER XXIII.

ANEMOMETRY; OR THE DIRECTION, FORCE, AND VELOCITY OF THE WIND.

THERE are such important differences between the fluid Ocean and the fluid Atmosphere, that the first impression on one's mind of *tides* caused directly by the moon, existing in both similarly, is soon effaced by reflection. At any rate, the only regular movement analogous to that of a tide, which instrumental means enable us to detect, is a diurnal (six hourly) change, which *appears* to be caused by the sun's influence,—whether by gravitation or by raising water in vapour, and the consequent electrical action, remains to be proved.

The time may come when even our theory of the tides, as caused by the moon's *attraction*, will be revised; meanwhile, however, it may well be asked—if the moon's *attraction* causes the tides or rise of the waters, why it does not do the same with the air? On the other hand, we cannot doubt some great cause in action coincident with the position and the phases of the moon, as observed from time immemorial, besides the phenomenon of the tides; for instance, the rise and fall of the sap in trees corresponding with the increase and decrease of the moon. The South American woodcutter never fells trees at the full—knowing that such timber will infallibly rot speedily through its moisture,—but waits for the utmost wane of the planet, when the sap is deficient. The same precaution is observed by the natives as to the use of stimulants and lowering medicines,—reserving the former for the period of the wane, and the latter for the increase and the full. Thus, the subject is well worthy of physiological and medical investigation. Are not inflammatory diseases invariably milder in the wane than at the full of the moon?[1]

As observed, however, in a previous page, the phases of the moon may be only the *time-measurers* of the effects in question, which may be due to compound planetary and solar causes. Such coincidences are nevertheless sufficient to uphold the popular belief, and the following estimate is the latest :—M. Kraszewski, of Romanow, has decided, from a long series of observations :—1. That 63 times in 100 the weather changes to bad when the moon crosses our equator. 2. When the moon's declination is north, bad weather occurs more frequently than when it is south. 3. When the moon is in its perigee, and at the equator, bad weather occurs 67 times in 100; when in its apogee, but still at the equator, there is bad weather 63 times in 100 ; and 4. Generally the moon near the equator determines

---

[1] The rise and fall of the sap of trees was observed by Mr. Evan Hopkins, by means of an appropriate section of the tree, exactly to correspond with the moon's age. When the moon is near the full or new, people are more irritable than at other times, and headaches and diseases of various kinds are worse. Insanity at these times has its worst paroxysms, and hence the origin of the term *lunacy*. The works of Drs. Meade, Sydenham, and Darwin abound with illustrations of this periodical influence, and Forster wrote a distinct treatise on the subject. (Forster, ' Atmospheric Diseases.')

bad weather in the same proportion. These coincidences, though very remarkable, are insufficient to establish the moon's influence as the direct cause of the
changes in question; but they should suffice to stimulate
inquiry on the part of astronomers, in order to discover
to what other causes the coincidences might be due.
Astronomers, following Arago, now condescend to consider the scientific import of our popular weather prognostics; why should they not direct their attention to
the moon also, universally held to be the arbitress of
the weather? And not only to the moon but to all the
planets of the solar system—all which must necessarily
influence the earth proportionately to their masses or
their distance. Why should they leave this legitimate
physical inquiry to the astrologers and so-called astro-
meteorologists, with their incomprehensible mysteries?
If all these sources of cosmical influence were accurately
studied, no doubt the missing links in the chain of wea-
ther-forecasting would be found and established.

The discrepancies in the results of the moon's supposed influence doubtless result, even with our present
information, from connecting them merely with her
phases, without taking into consideration the position
of the sun and her own position in her monthly progress,—in other words, the times of the equinoxes and
the solstices, the moon's declination north or south, her
perigee and apogee. As far as these grand meteorological epochs are concerned, we have no doubt whatever
that certain conclusions may be arrived at as to impending weather; but the explanation would require
more space than can be spared in the present work, and
must therefore be deferred to a future occasion.

Within the lower range of our atmosphere there are

varying strata—as shown by clouds and balloons; and
we may not only assume that there are variations also
above in the other thirty-five miles or more, but the
flight of *meteors* seem to demonstrate it conclusively.
We see them start, say, from the *south*, and, after running a few degrees, suddenly return, and proceed as
though they came from the *north*. They dart from the
S.E., and, after running some degrees, end as though
they came from N.E. Finally, they appear in the
N.N.E., and instead of describing a regular trajectory,
they vacillate and serpentine in their course; or coming
from the N. they end as though rushing from the S.W.
What can all these motions mean but the meeting of
disturbing forces? For, if there were no obstacle, the
meteors would always take the direction of the atmospheric layer from which they start. Hence, a method
of prognosticating coming wind, namely, by the re-
sultant of meteoric flights, as suggested, and apparently
demonstrated, by Coulvier Gravier.

Apart from its chemical composition, air is so different
from water in its variability, its great elasticity, and extreme mobility, and it is acted on daily to so great an
extent by the sun's influence, that we may perhaps en-
tertain doubts of the reality of *atmospheric waves*, as
generally understood, even while tracing their *apparent*
progress, and laying down undulations in diagrams.
That lines, or rather areas, of high and low barometer
succeed each other, and move eastward, has been proved
—to the great advantage of meteorology; and that the
curves of diagrams showing such alterations seem like
the profile outlines of waves, is obvious; but that these
changes are not exactly what they seem may still be
shown. Perhaps they indicate great pulsations, so to

speak, of the atmosphere, caused by alternate polar and equatorial action.[1]

It is abundantly evident that every wind makes a weather, and, consequently, it is of the utmost importance in meteorology that the direction of the wind at any given time should be accurately noted and recorded. All the various data which make up a probable weather prognostic are intimately connected with the direction of the wind, as explained at large in previous pages.

Besides the direction, however, it is most important to ascertain the force of a wind, not only because its duration is proportionate to its force, but also on account of the difference in the effect of a wind on *evaporation*, which so materially affects the atmosphere with respect to its moisture and electricity—the great agents in the production of the weather. Few imagine the difference which is made in this respect by the force of the wind. Suppose, for instance, the amount of evaporation from a vessel of six inches diameter to be 1·66 grains, at the temperature of 50°. Then, with a moderate breeze the evaporation will increase to 2·06 grains; and in a high wind to 2·51 grains.

Therefore, knowing the force of the wind, and provided with a table such as Daniell has drawn up, giving the evaporation of each force, we may ascertain the force of evaporation at the existing state of the atmosphere, after finding the dew-point by the hygrometer.[2]

This is a most important matter to horticulturists and to farmers. The amount of evaporation from the soil, and of exhalation from the foliage of the vegetable kingdom, depends upon two circumstances,—the satura-

---

[1] FitzRoy.   [2] See p. 61.

tion of the air with moisture, and the velocity of its motion, or the force of the wind. They are in inverse proportion to the former, and in direct proportion to the latter. When the air is dry, vapour ascends in it with great rapidity from every surface capable of affording it, and the energy of this action is greatly promoted by wind, which removes it from the exhaling body as fast as it is formed, and prevents that accumulation which would otherwise arrest the process; for the quantity of vapour that can be held by the air at any time is definite, and all evaporation ceases when that point is attained.

Now, over the saturation of the air with moisture the horticulturist has little or no control in the open air, but over its velocity, or " the wind," he has some command; and it is important that he should know when the force is such as to require " breaking," which he can effect by artificial means,—such as walls, palings, hedges, or other screens; or he may find natural shelter in situations upon the acclivities of hills. Excessive exhalation is very injurious to many of the processes of vegetation, and no slight proportion of what is commonly called *blight* may be attributed to this cause. Evaporation increases in a prodigiously rapid ratio with the velocity of the wind, and anything which retards the motion of the latter is very efficacious in diminishing the amount of the former. The same surface which, in a calm state of the air, would exhale 100 parts of moisture, would yield 125 in a " moderate breeze," and 150 in a " high wind."

The following table shows the rate of velocity per hour, and the force of the winds on a square foot of surface, according to their common designations :—

Common Name.	Rate per hour.	Force in pounds.
A great hurricane .	120 miles . . .	58 lb.
A hurricane . . .	88 ,, . . .	31·25 lb.
A storm	62 ,, . . .	15·62 ,,
A high wind . . .	36 ,, . . .	5.20 ,,
A brisk gale . . .	16 ,, . . .	1·10 ,,
A pleasant wind . .	8 ,, . . .	0·26 ,,
A gentle breeze . .	4·5 ,, . . .	0·07 ,,

Of course this table merely gives us a rough idea of the velocity and force of the wind, serving to account for the prodigious effects of hurricanes in the tropics and gales in the higher latitudes. A West India hurricane reached Newfoundland—3000 miles off—in six days. This, however, is only at the rate of about twenty-one miles per hour; but it must be stated that the north-ward motion of hurricanes is very small, compared to their direct force in the path of their gyration or whirl-ing, by which the strongest buildings and the stoutest trees are levelled with the ground, men and animals are lifted into the air, and cannon (12-pounders) have been hurled to a distance of 400 feet.[1]

It appears that the average variation in the strength of the wind during the twenty-four hours is as follows :— 11 miles per hour, the minimum force, occurring at 1½ A.M.; until 6 A.M. it remains much the same, being then 11·3 miles per hour; at 10 A.M. it is 13·4 miles per hour; at 1½ P.M. the wind is at its maximum strength, being 14·8 miles per hour; at 5 P.M. it is again 13·4 miles per hour, and at 9 P.M. 11·3 miles per hour. Hence it appears that the wind falls to its minimum force more gradually than it rises to its maximum; that the decrease and increase are equal and contrary, so that the curve is symmetrical; and that generally the force of wind is less at night than during the day.

[1] Dové's 'Law of Storms.'

Again, as to the direction of the wind. The dryness of the atmosphere in spring renders the effect most injurious to the tender shoots of this season of the year, and the easterly winds especially are most to be opposed in their course. The moisture of the air flowing from any point between N.E. and S.E. inclusive, is to that of the air from the opposite quarter of the compass in the proportion of 814 to 907 upon the average of the whole year; and it is no uncommon thing in spring for the dew-point to be more than 20 degrees below the temperature of the atmosphere in the shade. Daniell found it to amount even to 30 degrees. The effect of such a degree of dryness is parching in the extreme; and if accompanied with *wind* is destructive to the blossoms of tender plants.

Such are the facts of the case, and therefore horticulturists are specially concerned in anemometry. They should know, not only the direction of the wind, but its force, since the latter is so intimately connected with the important effects which they have to prevent by all the means within their power.

Elaborate and costly instruments are not absolutely necessary for supplying gardeners and farmers with this information. Messrs. Negretti and Zambra are engaged in the construction of an anemometer, modified from a German design, expressly adapted for the purpose of showing both the direction and the force of wind by inspection, even at the distance of sixty feet below the instrument, placed as usual in the best position for accurate action.

Ascending from these particulars to the general bearings of the subject, anemometry presents itself as the meteorological field requiring the greatest cultivation. With the exception of their general causes, very little is known concerning the winds. And yet, were accurate observations of the winds more general and more accurately recorded, there can be no doubt that something like a satisfactory theory of the winds might be very soon established. Nothing is more curious than the tracing of a wind throughout its travels,—giving rise to the adage that " most winds are liars ;" they rarely blow from the point whence they originate.

Of course, differences of temperature are their causes, and they influence their directions throughout their pilgrimage. Suppose, for instance, that a general S.W. wind occupies the upper regions, but that the western part of Europe is very hot, while the eastern regions remain very cold, with a clouded sky. This difference of temperature will immediately give rise to an east wind ; and when this wind meets that from the S.W., there will be a S.E. wind, which may be transformed into a true south wind. These differences of temperature explain the existence of almost all winds. Now, suppose that a region is unusually heated, and that there is no prevailing wind; then the hot air will flow in on all sides, and, according as the observer is on the north, the east, the south, or the west, he will feel a different wind blowing from the corresponding points of the horizon. However, to put this fact beyond doubt, we need corresponding observations, embracing a great number of localities, provided with accurate anemometers.

In a previous Chapter (xiv. p. 186), we drew attention to the fact, that the rotation of the wind, "with the sun," or like the hands of a clock, indicates the character of the year's weather. There is a close relation between the periods of *maxima* and *minima* of direct motion, and good and bad years for the crops, as before explained.

It is obvious that in this important matter, self-recording anemometers, fixed up throughout the country, would be of the greatest advantage.

Finally, a fall of rain or snow is always more or less productive of wind currents whose direction it is important to verify by self-recording anemometers. A few remarks on this interesting subject may serve to guide inquiry.

The gales of wind from the south-west, so prevalent in mild winters, may be accounted for by the reduction of the specific gravity of the atmosphere in the distant regions north-east of us, owing to rain or snow, and according to the extent and rapidity will be the degree of violence which we experience from the passage of the air, which rushes on to restore the equilibrium. If a considerable precipitation of rain or snow takes place in Russia or Siberia, a stream of air will first be experienced in the Baltic, the barometer will fall, and the wind regularly commence in Sweden, Denmark, and on the north-east coast of England, extending thus in a south-westerly direction, the mercury continuing low, until the equilibrium is restored.

When the storm is over in these places it then frequently rises very suddenly; for, as soon as the equilibrium is restored, the neighbouring countries commonly experience an accumulation of air, occasioned by the motion of the stream, continuing *after the effect ceases*, from the law of moving bodies, called *vis inertiæ*, which causes them, when once put in motion, to proceed in the same line for some time after the impulsive causes have ceased to act.

The stream of air experienced in England when a precipitation of vapour happens in Russia or Siberia, is not confined to one point of the compass. Thus, supposing a copious and extensive precipitation from the atmosphere to happen in the northern parts of Russia, then will Sweden, Denmark, England, and Ireland, experience gales of wind from the south-west; in France, southwest; Germany, Italy, the Adriatic, and the Euxine seas, due south; Tartary and Kamtchatka, south-east and east; and from the Polar Regions, due north. This hypothesis is strengthened by the circumstance that we never experience violent gales from the *south-west* except in *open weather;* that is, we have no very high winds from *south-west* during a *settled frost* on the Continent.

Sometimes these gales of wind continue for a fortnight or three weeks, with alternate calms or gentle winds, attended with great fluctuations of the barometer.

After a fall of rain the barometer usually rises, and would always rise, if there was no precipitation in the vicinity to draw off a portion of air from the place of observation. This rise of the mercury after rain is caused by the air flowing in from adjacent places where the density is greater, and from the continual evaporation of water.

It is obvious that the Meteorologic Office, which may be informed of the occurrence of such phenomena by telegraph, should be able, on these principles, to warn us of impending gales with a degree of certainty far greater than has as yet been attained.

Evidently the great Atlantic Telegraph might be made to warn us of impending gales, by generously and reciprocally supplying meteorological data as to the great currents of the atmosphere. The consideration of the following facts will show the utility of this suggestion. Let us suppose for example, the northerly current to be prevailing in America, and the southerly one in Europe; the latter will have a tendency to swerve from the point of contact, and allow the former to encounter it as a north-west wind. If, on the other hand, the northerly current is in Europe, and the southerly one in America, the latter will have a tendency to impinge on that from the north in a more westerly direction than that of its original course. This effect will be caused by the greater density of the polar current in the upper regions of the atmosphere. If, however, the polar current makes too great a resistance, a whirlwind will be produced in the equatorial current in the direction S., E., N., W. In the event of the aerial currents actually penetrating each other, a shifting of the vane occurs in the direction S., W., N., E., which naturally has nothing in common with the effect produced by a whirlwind-storm.

Meteorological intelligence from one country to another is always desirable. When the barometer *falls* in a country, it is because the temperature of this country is higher than that of the neighbouring countries, either because it is heated directly or because these countries are cooled. On the contrary, the *rise* of the barometer proves that this country becomes colder than those which surround it. A great fall of the barometer or frequent oscillations of the column, prove that there are meteorological disturbances on the surface of the globe, and conflicts of opposite winds, which change the weather; further, when the barometer rises and falls rapidly it is certain that the weather will be variable for a long time. If we knew the weather that prevailed in the rest of the globe, we might prognosticate that which might be expected. We ought to know, when the barometer is low, whether the cold is intense in America or in Asia. If in America, then the west winds will bring rain; if in Asia, the east winds will bring cold.

In studying, in the spring, the barometer and the direction of the gales of wind, we may establish certain probabilities. Thus, if the barometer has fallen consider-ably during S.W. winds, and then rises *slowly*, if the wind passes from W. to N.W. and remains in that direction, it is a proof of the predominance of west winds, and the weather will be influenced by them; as occurred in 1833. If, on the contrary, the barometer rises *very quickly*, and if the wind passes in a short space of time from S.W. to N.E., where it stops, then we may expect a prolonged cold, like that which prevailed in 1829.

A great fall in the barometer, or indeed any fall, in a place, must be compensated by a proportionate rise in some other; it is obvious, therefore, that the more such intelligence is extended the greater will be our probabi-lities as to the knowledge of coming weather.

In connection with anemometry we must glance at our Wreck Register—to the diminution of whose figures practical meteorology is intensely directed.

Westerly gales are far more destructive to shipping than gales from any other quarter. The following sum-mary of the most fatal winds during the year 1865 may be taken as the general standard.

Winds.	Gales.	Winds.	Gales.
N.	61	S.	94
N.N.E.	59	S.S.W.	133
N.E.	90	S.W.	192
E.N.E.	58	W.S.W.	102
E.	55	W.	73
E.S.E.	56	W.N.W.	91
S.E.	97	N.W.	101
S.S.E.	60	N.N.W.	59
		Total	1381

2 B 2

Again, we find that distinguishing the casualties of the past seven years according to *the force of the wind* at the time at which they happened, 678 occurred when the wind was at force 6 or under, of the conventional standard, that is to say, when the force of the wind did not exceed a strong breeze, in which the ship could carry single reefs and top-gallant sails; and that only 810 happened with a wind at force 9 and upwards, that is to say, from a strong gale to a hurricane.

Hence, the utility of knowing the force of the wind, which will enable us to infer the casualties it is likely to cause on our coasts and its approaches. In the last seven years, 118 casualties took place in a calm; 176 in light air or just sufficient to give steerage way; 450 in light breeze; 220 in gentle breeze; 784 in moderate breeze; 1280 in fresh breeze; 1217 in strong breeze; 441 in moderate gale; 836 in fresh gale; 1873 in strong gale; 1444 in whole gale; 505 in a storm; 693 in a hurricane; 50 variable; and 400 unknown.

A foreigner, looking at the Wreck Chart of the British Isles, might not unnaturally conceive that a very large proportion of the ships that pass to and from our ports every year were wrecked on our shores. When, however, he came to be informed that the number of vessels that cleared outwards and entered inwards last year alone, from the different parts of the United Kingdom (without counting vessels employed solely as passenger ships) was 409,255,—that they represented a tonnage of 65,231,034,—and that the value of their cargoes must be estimated at not less than £500,000,000,—the said foreigner would probably be much surprised, after all, to learn that not *one per cent.* of this great multitude of vessels was wrecked either in our narrow seas or on our coasts. Such, however, are the facts of the case, and it is not for us to justify even the loss of this relatively small amount of valuable property.[1] On the contrary, we contend that as education advances, and careful and thoughtful habits are instilled into sailors, this percentage of wrecks must diminish,—especially if accurate elementary principles of meteorology can be put forth to enable navigators to prognosticate the storm with certainty, and to be able to avoid its greatest violence by the skilful shaping of their course at its first approach.

ANEMOMETERS.

D'Ons-en-Bray, a distinguished mechanician of the last century, appears to have been the first who constructed an instrument of this kind, namely, "an anemometer which marks itself on paper, not only the winds that have blown during the twenty-four hours, but also their different velocities or relative forces." (A.D. 1734.) This anemometer consisted of a vertical cylinder mounted on the same axis as the vane, carrying twenty-five pencils of equal lengths, planted perpendicularly to its axis, in the direction of a helix, forming a complete spiral, which was thus divided into twenty-four equal parts. A band of paper, by means of clockwork, was drawn parallel to itself, so as to be grazed slightly by one of the pencils, for a certain direction of the vane; and the cylinder, which followed all the movements of the latter, always brought to the surface of the paper one of the twenty-four pencils, whatever the direction of the wind. Thus, the paper, the whole of the surface of which passed successively before the cylinder, carried in the direction of its axis, a succession of pencil-marks, the height of which indicated the direction of the wind, and the length of which was proportionate to the time during which that wind had been blowing.[1]

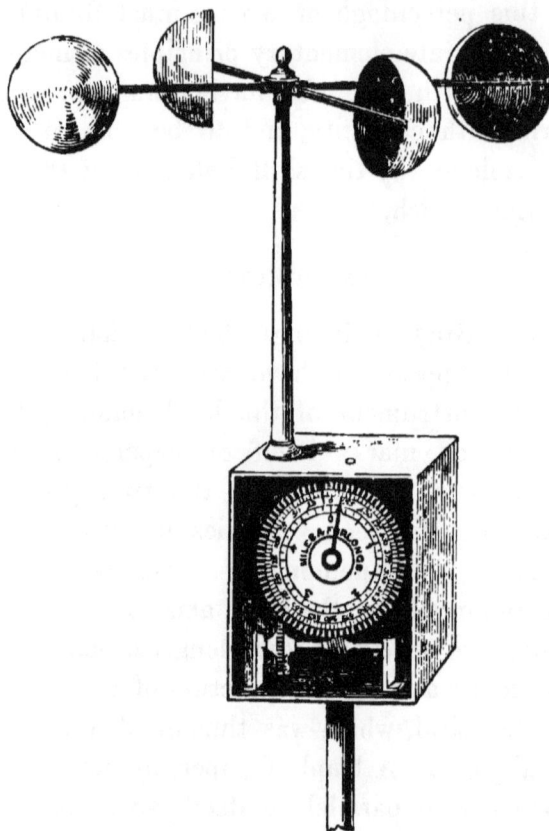

The Anemometer.

The general principle of our modern anemometer may be shown by the above figure of the one invented by Dr. Robinson, of Armagh, as shown in the figure, and improved by Casella. It consists of four arms, at

[1] Mém. de l'Ancienne Académie des Sciences (year 1734, p. 123).

the ends of which there are light hemispherical cups, having their diametral planes exposed to a passing current of air; they are carried by four horizontal arms attached to a vertical shaft or axis, which is caused to rotate by the *velocity* of the wind.   Dr. Robinson found that the cups, and consequently the axis to which they are attached, revolve with *one third the wind's velocity*. A simple arrangement of wheels and endless screws is appended to the instrument, which, by means of *two* indices, shows on inspection the space traversed by the wind.   To this arrangement has been added two additional indices, which extend the indications from 500 to 5,000 miles.

There are various other anemometers, such as Lind's, Dr. Whewell's, and Ostler's; two of the latter were sent by the Government, in 1859, to Bermuda and Halifax, Nova Scotia, for anemometrical observation, and some of the records have been published by the Board of Trade.

Unfortunately, up to the present time, the cost of a perfect instrument of the kind has prevented the general use of anemometers.   On the one hand, we have had the well-known and excellent, but very costly instrument of the British Association and the Royal Kew Observatory, the price of which was £70; and on the other, small ones of ordinary construction at from £3 to £4, registering velocity only, without showing either the time or direction.

Recently, however, two instruments have been contrived, one by Mr. Howlett, late of the War Department, the other by Mr. Casella, of Hatton Garden, which must be considered advances in anemometry.

The following is a figure of Howlett's ANEMOGRAPH,

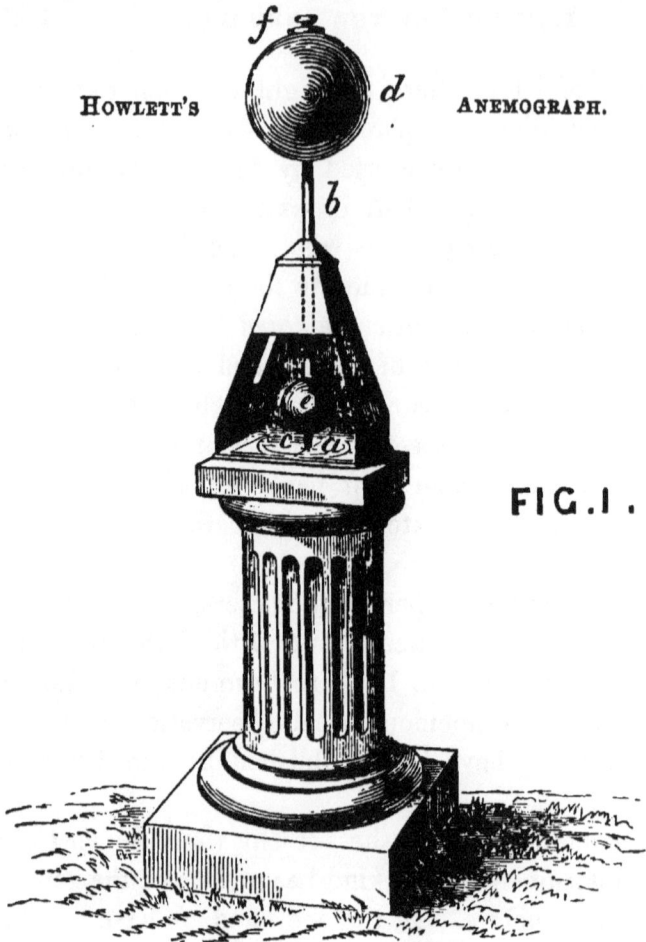

Howlett's   Anemograph.

FIG. I.

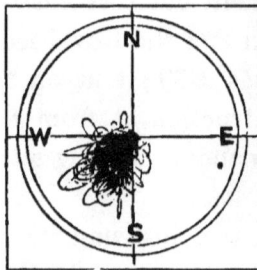

FIG. 2.

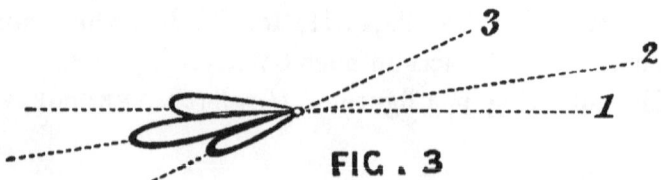

FIG. 3

introduced by Messrs. Elliott, Brothers, of Charing Cross, together with specimens of its performance as a self-recording instrument of the direction and the pressure of the wind.

Fig. 1 is a view of the instrument. The base *a* is a slate, on which is engraved a circle 10 inches in diameter, divided into degrees, and figured from 0 to 360. Upon this base is fixed a square pyramid made of zinc, having a window on each side, and closed by a shutter. A brass tube *b*, forming a lever, works in a gimbal as a fulcrum in the top of the pyramid. A pencil or tracer, *c*, works freely in the brass tube. The sphere *d* is capable of being moved up and down, and is of such a diameter that the pressure of the wind on its hemisphere shall be equal to the whole or any required portion of a square foot; and a weight, *e*, is so adjusted with reference to the sphere, as to cause the pencil to express pounds of pressure on the square foot by its distance from the centre of the graduated circle accordingly as the sphere is down, up, or up with a weight at the top, thus giving three scales, namely, from 0 to 20 lbs. on the square foot,—from 0 to 5 lbs.—and to 2½ lbs. A wooden measure, graduated to these three scales by actual trials, is attached to the instrument.

In action, the pencil throws out from the centre, or zero, a line in the direction the wind comes from, and, in returning, a *loop* or curved line is formed; and the force of the wind is indicated by the length of the line or loop; so that by laying the wooden scale against the centre of the slate and the end of the loop, the force is read on the edge in pounds pressure on the square foot, and the angle at which the wind crossed the meridian is at the same time found in degrees on the divided scale.

Thus, the instrument records the action of the wind in the form of a map. The part shown black, Fig. 2, represents the manner in which either a breeze or a storm is recorded, the salient points of which mark the direction and force of the principal currents, measured by applying the scale before mentioned. There is a contrivance for checking oscillation.

The *loops* formed by the pencil of the anemograph draw attention to an important meteorological fact. In all winds, if the vane or weathercock be observed, it will be found that the latter deviates, more or less, to one side or the other, from the prevailing wind. The side to which it deviates will infallibly be the direction of the next change or shift of wind. This is important in the prognostication of the weather; and, on examining a great many of the daily wind-maps traced by this instrument, we found this fact completely established; so that we should be able, by mere inspection of the performance of this instrument, to ascertain the next direction of the wind—which piece of information may be reasoned out to a very probable prognostication of coming weather.

We believe these loops to be the necessary product of another current or other currents of air besides the prevailing wind. At the same time, we feel bound to state that the ingenious inventor of the instrument believes that these loops have a different significance. He says, "If, while the wind is blowing, the pencil be held on the centre of the slate, and then let go for a few seconds only, we get a figure consisting of several loops, as shown by the dark lines,—from which, probably, no other conclusion can be drawn, than that the wind moves on in *circles* which are constantly crossing the paths of each other, as shown by the lines 1, 2, and 3." We may be permitted to doubt this conclusion. *All* the tracings of the pencil are loops, whether the pencil be held or not, and the arresting of it can obviously have no other effect than that of oscillation, owing to two forces acting upon it. It seems that our explanation of the loops is supported by the difference in the shape of the loops traced by the leading winds. The loops are very wide during the prevalence of the equatorial S.W., less during the polar N.E., and least of all during an east wind, when the curve approximates the shape of a very acute angle. It is scarcely necessary to state that the play of the currents which constitute the respective winds is restricted analogously to the difference of the curves or loops traced by the faithful pencil of the anemograph in its diurnal variation.

This instrument is a decided advance in anemometry, not only in being very portable and not at all costly, but also by the nature of its descriptive indications of the winds, as it were analysing them on all occasions, and by the width of the loops furnishing some idea of the width of country swept by the respective currents. It is obvious that deductions drawn from the study of such elements may be turned to account in prognosticating consequent weather; because by knowing the nature of the band or zone of country over which any wind blows we can estimate the modified effects from the configuration and nature of such surface—in a word, solve the problem of all the actions and reactions which make up every wind that blows. Still further to aid such a study, each of the daily maps of the wind should have recorded upon it the barometric and thermometric means, the mean humidity of the twenty-four hours, and the rainfall or snowfall, if any. With such a daily map before us, there should be no difficulty in prognosticating the consequent weather of at least twenty-four hours; by studying the *month* we may get a clue to the weather of the subsequent month; and by studying the maps of a season form a very probable opinion as to that which is to follow. We apprehend that the ane-mograph might be made to register the time as well as the direction and force, without great addition to the cost of the instrument.

Casella's self-recording embossing anemometer (see FRONTISPIECE) may be compared to that of the Royal Kew Observatory in its capabilities, but differing vastly in cost. In its construction Mr. Casella was aided by Mr. Beckley, the engineer of the Kew Observatory. The instrument was exhibited and a paper read on it at the last meeting of the British Association, and it may be described as follows :—

The size is about one-third, and the weight one-fourth of its larger predecessor, whilst the price is less than one-half.

The mode of action is equally simplified.

Thus, the cups revolving turn a pair of embossing rollers, through which they draw a narrow strip of paper and impress it with figures, which show the rate at which the wind has been travelling. One revolution of these rollers represents fifty miles of horizontal move-ment of wind.

The direction is shown on the same slip by means of a small arrow, which is turned by the vane—their move-ments being identical; and a small hammer, moved by the clock, is thus made to impress the direction at every hour, or at shorter intervals if preferred.

Another important advantage obtained by using this anemometer is, that the paper need only be changed, and observations made, once a week, a month, or even at longer intervals, whilst the cost of the narrow strips of paper employed is less than one-tenth of the usual charge.

It is impossible to overrate the value of this ingenious modification of the larger and more expensive instrument. This anemometer might be placed on each townhall in the kingdom for the advancement of meteorology : the simplicity of its arrangement and durability, with the little care required in its use, render now easy that which has hitherto been very difficult owing to the labour, cost, and uncertainty which have attended the fixing of the larger instrument.

It is evident that self-recording instruments are urgently needed in the present state of meteorological science, and that they will soon, in all probability, be largely employed both in this country and abroad. The Royal Society have recommended the establishment in the British Isles of six stations with self-recording instruments, for the purpose of making and recording full, accurate, and *continuous* observations of meteorological phenomena at those stations. The advantages of such instruments are manifest. By reason of the continuity of their records, no wave or variation of any description in any of the meteorological elements can escape notice, and the course of that wave or variation can be tracked with certainty from station to station, and its modification at the time of reaching each station in succession can be accurately observed. For the same reason one difficulty now seriously felt in charting the weather, namely, that which arises from observers in different

places and countries adopting different hours of observation, would wholly disappear; and a further difficulty
—that which arises from observers being unpunctual to
their professed hours of observation, would also disappear. The unvarying accuracy of the record is an
advantage of still greater importance than might be
expected by those who have no experience of the frequent errors to be found in meteorological registers.
Each error creates considerable confusion; it throws
doubt on the observations accurately made at neighbouring places; and that doubt cannot be removed except
by the continuity of the records at those places. Another advantage of self-recording instruments is that
their records are independent of particular scales. Their
notation is in lines and curves, which can be measured
with equal facility according to any desired scale. The
thermometer lines could be measured at pleasure according to Fahrenheit's scale, as used in England,—to the
Centigrade, as in France,—or to Réaumur's, as in Germany. The barometer lines could be measured with
equal ease in English inches, in millimetres, or in Paris
feet. For these various reasons, self-recording instruments are of eminent local and international utility.

But it is not sufficient for the purpose to observe the
weather of the British Isles alone. The experience of
meteorologists, abundantly illustrated by the daily weather maps of M. Leverrier, show, beyond all doubt,
that the weather-changes of England, and even of all
Europe, are but parts of immense systems. These systems reach southward to the trade-winds, and with them
far in the direction of the Gulf of Mexico, whilst they
are of unknown extent to the north. The area of the
North Atlantic, and especially of the Gulf Stream,

appears to exercise a most important influence on the generation of the storms and weather-changes that affect England. Under a conviction of the importance of studying the weather on a sufficiently extended basis, Leverrier is now engaged in producing charts of the northern hemisphere, between the equator and lat. 70° N., and long. 100° W. and 60° E., for each day of the year 1864. What an instructive meteorological study these charts will form when completed! Doubtless, this country should take a share in inquiries of this description, proportionate to her means of obtaining information.

The "Remarks" daily published by the Meteorologic Office respecting "areas" of pressure should be more precise, and always accompanied by the statement of the prevailing *wind*. An area of *low* pressure at Vienna, as before stated, would signify an *east* wind; at St. Petersburg, a *north-west* wind; and in Holland it appears that, whilst a high reading of the barometer at the northern stations is followed by an easterly wind, the same reading at southern stations is followed by a wind from the westward. Obviously, these are important data in drawing conclusions as to coming weather form the meteorological facts around us or in the distance.

In theoretical as well as practical meteorology we hold that every hint, however trivial in appearance, should be duly considered. The most vulgar prejudices may originate important theories,—as, for instance, that which connects the periodic swarms of meteors with the perturbations of temperature, as demonstrated by M. Deville. (See Chapter XIX.)

There is a vulgar prejudice which has prevailed from

time immemorial in Sussex, that a *Saturday's moon*
brings blowing and wet weather. By some accident
this proved very true during twenty years of Forster's
meteorological experience. To ascribe such a phenomenon to the occurrence of the new moon on the day
dedicated to Saturn must, of course, appear supersti-
tious; but there may be natural causes why the conjunction of the sun and moon, happening at some such
diurnal periods, may, in the long-run, turn out to be
connected with rough weather; and these periods once
falling on a Saturday, would for a long time continue to ·
do so,—hence may have arisen this vulgar notion. Old
shepherds, gardeners, hunters, and men of education,
have alike testified to the fact.[1]

*Indeed, the whole doctrine of periodic phenomena is
very little understood.*

In connection with this important subject, we venture to put forth some suggestions kindly communicated
to us by Mr. Fullbrook, the well-known writer on rainfall and meteorology. From extensive observations he
believes he has established the fact that, whenever two
planets form a right line with the earth, some disturbance connected with their light, electricity, or magnetic
reaction takes place in the atmosphere, and a much
larger quantity of rain—especially about the fourth or
fifth day after—is the result, and the greater number of
the more violent and extensive storms and hurricanes have
occurred at or about the time of this *excess* of rain.[2] To

---

[1] Forster, 'Prognostics of the Weather.'

[2] This fact, Mr. Fullbrook assures us he has verified by the investigation of 300 consecutive conjunctions of the planets in longitude, using
the registers of Luke Howard for the vicinity of London, between the
years 1807 and 1830 inclusive. Passing over small irregularities, which

the effect of rainfall and snowfall, as causing atmospheric disturbance, we have before alluded (p. 362) ; and, as Mr. Fullbrook observes, there is nothing, perhaps, so much calculated to impress ordinary minds with the idea of uncertainty in the weather as the occurrence, during a rainy season, of a calm, clear, warm day, whose lovely sky inspires the hope that many more will follow in its train ; but which is on the morrow—perhaps sooner—obscured by wild-looking clouds, which presently become dense and dark, and pour down torrents of rain, perhaps accompanied by thunder and lightning. The wind blows up first from the south, then veers to the south-west and west, and is found in a day or two blowing from the north-west or north, with a great reduction of temperature, and probably some indication of frost in the early morn. Now every one we meet exclaims, " How changeable and uncertain the weather is !" Strange, however, as it may appear, these *various* effects, so suggestive of chance to our finite minds, have all arisen in their proper order of succession from one definite cause ; they exhibit the different stages of · a process wisely designed to bring about a bountiful distribution of rain over the face of the country, to water the numerous productions of the soil, and to replenish our springs and fountains.

This process may be described as follows :—

1. The atmosphere resting on some limited portion of,

will always exist unless the investigation is extended over an extraordinarily long period, the quantities of rain increase for some days until the fifth day after the conjunction, when there fell no less than 29·11 inches in the 300 days (being one day to each of the 300 conjunctions),—the average at London being only 25¼ inches for 365 days. The amount ' then decreases during several days.

the globe, which from its position or state at the time is more exposed to, or susceptible of certain external influences,—as those before alluded to with respect to the other planets of the solar system,—has its temperature thereby considerably raised, a condition so often noticed before storms and heavy rains.

2. This increase of temperature dissipates all clouds and mists at the time floating in the atmosphere, and causes a temporary suspension of the rain (if any) ; hence, the brief interval of fine calm weather, and the transient clearness before rain comes on again, which renders distant hills and objects unusually visible.   Surface evaporation is also stimulated, and hence exposed surfaces dry up more rapidly before rain.

3. The increase of heat also rarefies the atmosphere, which, expanding upwards, overflows from its summit, the excess going off to the surrounding regions ; hence, the central atmospheric column becomes lighter, and it is rendered mechanically still more so by the larger proportion of aqueous vapour which is now infused into it and which is lighter than common air.   Owing to these causes, the column becomes much lighter, as indicated by the rapid fall of the barometer.

4. The atmosphere now pressing less on the surface than before, evaporation proceeds more rapidly,—a fact demonstrated by experiment.   And here must be noticed a wonderful arrangement of Providence.   In cold climates, and also in England during the winter season, when evaporation is slow in consequence of the low temperature, a greater *reduction of pressure* takes place on these occasions, which makes up in a great measure for the more feeble action of heat in evaporating moist surfaces.

5. The lighter atmospheric column induces an indraught of air at the surface from the surrounding regions, whose atmosphere has become somewhat heavier in consequence of the previous outflowing to their summits; and here we see one cause of winds blowing from opposite quarters in connection with storms and heavy falls of rain.

6. The atmosphere at length becomes charged with the aqueous vapour which has been thrown up from the surface, and also brought in from the surrounding districts; and that which has ascended to the higher and colder aerial strata begins to condense into *cloud*, which gradually increases in density, and descends, assuming those various appearances so highly indicative of approaching rain. And here may be mentioned the probable cause of one type of cloud, so highly ominous of heavy weather at hand. It is well known that there often exist currents in the atmosphere flowing under each other in different directions, and, as it were, brushing each other's surfaces. Now, a sheet of cloud formed in an upper current will have the motion of such current, and will of course retain it when, through an increase of weight, it descends on the surface of the next current; as this is moving in a different direction, it acts on the under surface of the cloud in the same way as wind acts on the surface of the sea, and gives it a curious *wild, wavy appearance.*

7. At length, dense clouds cover the sky, which produce a quantity of rain. And now another potent cause of wind comes into operation. When invisible vapour is condensed into rain, it occupies a space *sixteen or seventeen hundred times less than it did before;* hence a considerable *vacuum* or *tenuity* occurs in the

2 c 2

air whenever large quantities of rain descend, and the air rushes in from the surrounding regions to fill the void thus occasioned, as previously explained (p. 362) ; and it is probably to this cause, acting in concert with the rarefaction due to heat before explained, that we must attribute those terrific hurricanes which sometimes sweep over the surface of our globe.

8. The usual rise of the barometer after rain plainly demonstrates the inflowing of air from surrounding regions in consequence of the above causes, for the conversion of vapour into rain, having partially or wholly ceased to diminish the volume, the influx of air fills up the vacuum and restores the equilibrium. (See p. 369).

9. Diminished atmospheric pressure must be the consequence elsewhere; and having extended to a considerable distance around, where fair weather probably continues, evaporation is there going on, and converting a comparatively small quantity of water into a *large* volume of vapour; it is, in fact, not only producing a supply for the heavy precipitation at the centre of the storm, but it is, in a manner, supplying the wind for its influx. Now, as evaporation will proceed more rapidly, and produce a larger volume of vapour, in the warmer latitudes and from the surface of the seas, it follows that the stronger winds will come from that quarter, and drive the storm in the opposite direction.   Here we see the reason why great storms, first arising in the tropical regions, in consequence of the greater heat and more abundant vapour there, proceed thence towards the poles,—not, indeed in a straight course, as the easterly wind of the torrid zone first takes them westward, and afterwards the greater prevalence of the west wind with us bends their course to the eastward.   The more

powerful south-west wind, which at such times sweeps across the British Isles, and which, for the reason above assigned, is due to our position, very much alters the character of the storm with us; yet every one will have noticed, in even small storms and gales, the tendency of the wind to shift round from south-east to west and north during the time they are traversing our island from west to east, and presenting the different parts of the cyclone to us in succession.

Agriculturists and others, whose operations are much affected by changes of weather, would do well to observe that, if rain and clouds quickly pass away and are soon succeeded by warmth and a clear day, with the wind between south and west, the change should not be trusted —it is only the prelude to a speedy return of more rain; and if the barometer rises too rapidly after rain, it is probably owing to the suspension mentioned (section 2), and it will as quickly fall again, introducing more wet and cloudy weather.

Such is Mr. Fullbrook's theory of storms and gales, and we doubt not that it will be appreciated as a useful summary of atmospheric causes and effects, more or less applicable on every occasion.

CHAPTER XXIV.

WHAT BECOMES OF THE SUNSHINE?

WHAT is the sun? has long been a debated question;
but now—thanks to the revelations of what is called
the *spectrum analysis*—we are in position to study
the chemical composition, not only of the sun, but
the most distant stars and nebulæ. In this respect,
man's power is no longer limited. He is no longer confined to his little sphere. He may now extend his researches throughout infinitude. The moment any star
sends us its light, we can discover what elements may
exist in it analogous to those in our planet, and get an
idea of its physical constitution. This seems incredible,
and yet the fact is as clearly established as any in the
domain of science; we know positively that there exist
in the sun's atmosphere at least nine of our metals,
namely, iron, sodium, magnesium, calcium, chromium,
nickel, barium, copper, zinc, but apparently no silver
nor gold; so that, although deserving every other good
and grand epithet, it seems, as yet at all events, that
the sun is not " golden." The gas hydrogen also exists
in the sun. In like manner, many of the substances
known on our earth have been detected in the atmo-

sphere of the stars by Mr. Huggins and Professor W. A. Miller, to whom we are indebted for this most important discovery. Thus, the star Aldebaran contains hydrogen, sodium, magnesium, calcium, iron, tellurium, antimony, bismuth, and mercury; whilst in Sirius only sodium, magnesium, and hydrogen have with certainty been detected.[1]

The branch of science which has led to these discoveries has been recently developed chiefly by the researches of Bunsen and Kirchhoff. It is founded on the fact that when chemical substances are strongly heated in the blowpipe or other colourless flame, they produce a peculiar colour, by the occurrence of which the presence of the substances themselves may be detected. The instrument used for the purpose is called the *spectroscope*, the various appearances from different substances being termed *spectra*. It will give some idea of the extreme delicacy of this mode of analysis if we state that a portion of common table-salt less than the one hundred and eighty millionth part of a grain can be detected. By means of it compounds are found to be most widely disseminated throughout the earth which were supposed to occur very seldom.

Now, when sunlight is allowed to fall on the slit of the spectroscope, it is found that the solar spectrum thus obtained differs essentially from the spectra to which we have alluded. The solar spectrum consists of a band of bright light, passing from red to violet, but intersected by a very large number of *fine black lines*, of different degrees of breadth and shade, which are always present, and always occupy exactly the same relative position in the solar spectrum.

[1] Roscoe, 'Elementary Chemistry.'

It is the study of these *dark lines* which led to the discovery of solar and stellar chemistry. On comparing their positions in the solar spectrum to those of the *bright* lines in the spectra of certain metals as produced in the spectroscope, it is found that each of the *bright* lines of any metal coincides, not only in position, but also in breadth and intensity, with a *dark* line in the solar spectrum; and if a solar and metallic spectrum be both allowed to fall, one below the other in the instrument, the bright lines of the metal are *all* seen to be continued in the *dark* lines of the solar spectrum. Hence the conclusion arrived at that there must be some connection between the bright lines of these metals and the coincident dark lines of the solar spectrum, as such coincidences cannot be the result of mere chance. But, indeed, the bright lines of the metallic spectrum may be reversed or changed into dark lines by allowing the rays of a strong white artificial light to pass through a flame coloured by soda, for instance, and then to fall upon the slit of the spectroscope. This experiment, applied to other substances, exhibits the same result; the inevitable conclusion, therefore is, that the dark lines in the solar spectrum are caused by the passage of white light through the *glowing vapour of the metals in question*, present in the sun's atmosphere. The sun's atmosphere, therefore, contains these metals in the condition of glowing gases—the white light proceeding from the solid or liquid, strongly heated mass of the sun which lies in the interior.[1]

The similarity of the substances of which these meteors

[1] Roscoe, *ubi suprà*. During the recent display of November meteors, a friend of Mr. Huggins endeavoured to apply the spectrum analysis to those momentary gleamings, and thought that he detected the

consist to those that constitute our earth plainly shows a community of origin ; but the presence of *carbon* in some of them has puzzled philosophers how to account for it, as this substance cannot exist excepting where there has been vegetable or animal matter. But perhaps the presence of carbon suggests the fact that these fragmentary planets are but the splinters of one or more large planets rent asunder, which, like our earth, had been clothed with vegetation and inhabited by animals. At all events, there seems to be no other way of accounting for the carbon found in these bodies.

Such are the salient facts of this new science, which is destined to throw great light, if we may use the phrase, on sunshine,—explaining the innumerable effects of light throughout creation and the arts of life. We can no longer consider light as merely consisting of infinitesimal *particles,* or as infinitesimal *waves ;* we may now conclude that it is *metallic,*—that sunshine consists of a metallic " shower," such as the Greek mythology ascribed to Jupiter in one of his unions with mortals (strange coincidence of fact with emblematic fiction!), but, instead of a shower of gold, according to the figment, the beneficent sunshine bathes us with elementary iron, sodium, magnesium, calcium, chromium, nickel, barium, copper, zinc, and hydrogen ! It is Jupiter, not

presence of the metal sodium in their light. Mr. A. Herschel had been impressed with the same conviction on a previous occasion.

In Forster's works we find a striking passage which may be taken as the first glimpse of spectrum analysis. He wrote some fifty years ago. " We may imitate the different colours of the spectra of the several stars and planets, by burning antimony, steel, and other metallic filings, in pyrotechnical jerbs, and viewing them through a prism. Compare the rismatic spectrum of ignited *steel* with that of Jupiter, of burning *timony* with Sirius, of *copper* filings with the spectrum of Mars, and on." (' Signs of the Seasons.')

Apollo that the ancient mythology should have identified with the sun. In the idea of *Zeus* (or the "Burner") and the *Zeuspater* (Father Jove) of the Latins, we might translate all the amours of the Father of the Gods into physical facts connected with the "arch-chemic sun," as Dante calls him, and recognize the idea that all animals are children of the sun. For let us consider the facts of the interesting inquiry. Animals cannot produce the complicated chemical compounds which they need for their structure; it is otherwise, however, with plants, which are able to build up their various parts from the elementary constituents. Now, this function of plants is entirely dependent upon sunlight—upon sunshine. Without sunlight the green colouring matter of the leaves of plants cannot decompose the atmospheric carbonic acid ; and therefore, without sunlight, the plants cannot grow. In order to separate the atoms of carbon and oxygen an expenditure of force is necessary, and this force is derived from the rapidly-vibrating solar rays. It is the sunbeams that tear asunder the carbon and oxygen atoms, and thus enable the leaves to take up and assimilate the carbon, throwing out the oxygen into the air for the subsequent use of animals, if for no compensating cosmical effect in the bargain.

Again, when vegetable matter is ignited, it burns, and in burning forms carbonic acid,—generating exactly the same amount of *force* as the vibrations of heat which were needed, in the form of vibrations of sunlight, originally to decompose the atmospheric carbonic acid. Hence, when coal burns, the light and heat evolved may truly be said to be that of the sun, for has not the sunshine made the carbon? But carbon is the solid part of fuel ; and it abounds also in all animal bodies, earths,

and in some minerals,—and nowhere without being referable to sunshine actually forming it at the present
time, or as having formed it in other epochs of creation.
Thus, then, as vegetables cannot exist without sunshine,
and animals cannot exist without vegetables, all animals
may be truly called, as we said, the children of the sun.

But if the sun be the source of life, he is also the *destroyer*, as seems to be signified by the Greek name and
personification, *Apollo*,[1] tending ever to effect incessant
change or "metamorphosis" throughout creation, in
which, scientifically and philosophically speaking, there
is no *death*, but merely change of state, the activity of
new vitalities tending either to restore elementary forms
or to effect new combinations; "forms changed into new
bodies," as Ovid happily hits off the economy of nature.[2]

Thus viewed, corruption, putrefaction, and decay cease
to be revolting. They are Nature's mode of operation,
by analysis and synthesis. Thus she fashions the most
perfect specimens of animal and vegetable beauty. Elements are indestructible. Succeeding epochs enjoy only
their usufruct or use. The plants of to-day live by the
same elements that supported the plants millions of ages
ago: it is the same with the animals of our epoch : every
particle of the matter we "assimilate" has done service
"over and over again" in the laboratory of Nature,—
under the influence of sunshine, —the Photosphere of the
Sun.

But the sun is also a destroyer. With the mighty
Photosphere of the Sun must also be connected the
causes of disease. Already is a reasonable theory ad-

---

[1] Απολλύω, "I destroy."

[2] "In nova fert animus mutatas discere formas
　　Corpora." 　　　　　*Ovid, Metamorph.*

vanced connecting *Terrestrial Magnetism* with *Epidemics of Fevers*. The facts of these periodical aggravations of fever being accompanied by a disturbed state of the earth's magnetism, and of their being apparently connected with unusual disturbance of the sun's photosphere, open up a field of inquiry of the deepest interest, which bids fair to throw into the shade the wildest fancies of the *astrologers*.

Magnetic storms, as they are called, seem in some way connected with the intensity of epidemics. These storms consist of great and rapid alterations of the force of the magnetic elements, causing the instruments to oscillate to a much greater extent, and in a more irregular manner than is usual for the hour of the day, or the season of the year. It has been observed by Sabine, that these storms recur in a regular period of about ten years, the minimum occurring in the years 3 and 4 in each decade, and the maximum in the years 8, 9, and 10, while the transitions from the minima to maxima, and *vice versâ*, are rather abrupt. Sabine has further drawn attention to the fact of these magnetic storms coinciding in period and epoch with the frequency of the *solar spots*. Those who have examined the records of epidemics may recollect that the years 7, 8, and 9, in each decade in the present century, have been characterized by an unusual prevalence and severity of fever in many parts of the world. Severe fevers are occasionally met with in the years 3, 4, and 5 of the decade, but they do not seem so diffused as in those numbered 7, 8, and 9. These details show a connection between magnetic excitement and the periodic aggravations of febrile diseases throughout a large portion of the world, more especially in the years 7, 8, and 9 of the decade ; but they do not indicate the steps in the conversion (to use an expression more in common use) of the magnetic force into *the active cause of disease,* and until more is known of the laws of terrestrial magnetism, and of the immediate causes of magnetic disturbances, we shall not be in a position to do more than point out their connection in a general manner, though, even here, some facts have been ascertained which promise a rich harvest if followed up.[1]

But we must hasten to leave this gloomy view for " the sunny side " of the subject. Without the light of the sun, plants cannot grow. The living germ, the green leaf, owe to the sun their power of transforming earthly elements into living, vigorous structures. The germ may, indeed, be evolved underground without the action of light, if provided with the adequate heat of sunshine; but only when it breaks through the surface of the soil does it first acquire the power, by the sun's rays, of converting inorganic elements into its own structure. The illuminating and heating rays of the sun in thus bestowing life, lose their own light and heat. Their power now becomes latent in the new products of the frame, which have been produced under their influence from carbonic acid, water, and ammonia.[2]

All the artificial light we use, whether from grease, oil, spirit, or paraffin, is still a continuation, although imperfect, of solar light—of sunshine; and the flaming gas from millions of burners that strive to change the night of London and other cities into day, is but the resurrection of primeval sunshine from the carbon of coal, which was once the wood of forests in " the far backward and abysm of time."

[1] 'Further Observations on *Pandemic Waves.*' By R. Lawsen, Esq., Deputy Inspector-General of Hospitals.   [2] Liebig.

This subject of coal and the coal-beds seems to demand a passing consideration. At the last meeting of the British Association for the Advancement of Science, Mr. Grove expressed himself as follows:—" At a moment when the prospective exhaustion of our coalfields, somewhat prematurely perhaps, occupies men's minds, there is much encouragement to be derived from the knowledge that we could at will produce heat by the expenditure of other forces."

The enthusiastic anticipations and promises of scientific men are always cheering and pleasant to hear, and in nothing more so than the vital matter involved in the present question. But, can we accept this pleasant prospect of science in this all-important deficiency with which we are threatened?

The basis of Mr. Grove's hope is the doctrine of the mutual convertibility of various forces, and it involves the supposition that man can *create* a force. This doctrine is not established on any satisfactory evidence. Man has no power to create even an infinitesimal quantity of force. The creation of force is the sole prerogative of the Great First Cause—of God Almighty. All man's power is strictly limited to the application of pre-existing forces by the action of his muscles, under the direction of his brain; and the exercise of his muscles, as well as the function of his brain, is wholly dependent upon the evolution of force from his *food*. *That* force is matter which has been elaborated under the influence of *solar light*, or, in other words, it represents a store of force derived exclusively from the sun, from sunshine. Although the food of man consists partly of vegetable and partly of animal substances, yet, as before shown, both owe their existence directly or indirectly to that of

light.   Purely carnivorous animals form no exception to
this universal law, for if they do not eat vegetables they
prey upon other animals which are herbivorous.   Only
let the sun be extinguished, and every living thing or
being must speedily die.

As these are incontrovertible facts, it follows that man
is, as it were, only a vehicle of pre-existent physical
force, to be applied as his will, well-directed or perverse,
may determine.   However humiliating this view may be
to proud spirits, it is nevertheless undeniably true.   Pride
must be silent when universal nature reads its awful les-
son.   The black thing which we call coal may be dug
out of the earth by the expenditure of an amount of
muscular force, excessively minute compared with the
force which it is capable of evolving when burnt.   The
force is in the coal, and coal is, so to speak, an accumulation of Sun Force, inasmuch as all coal may be shown
to have been derived from plants, and all plants from
solar influence.   Neither Mr. Grove nor any one else can
be justified in hazarding even a conjecture as to the
future probable discovery of a *substitute* for coal; the
thing is an impossibility, and the "encouragement" of
which the learned Professor speaks must be regarded as
groundless and delusive, not to say mischievous.[1]

---

[1] It has been said that so long as its people are industrious and reso-
lute, Great Britain will be the highway and the mart of nations; but,
supposing our manufactures greatly to decline, as inevitably will be the
case when our coal either is exhausted, or what is equivalent, has become much dearer than the coal of other manufacturing nations,—to
what, it may be asked, is the industry of our people to be devoted, with
all their resolution?   Not to agriculture exclusively, for our population
at present is greatly in excess of what is needed for the cultivation of
the land, and, besides, the products of our agriculture can never compete
with those supplied by that of the foreigner.   What is happening at
this moment in Cornwall from the exhaustion of its tin and copper will

With respect to the threatened deficiency of coal, all that can be suggested is the utmost economy in its use, adequate means being taken to stop the waste of the precious material by preventing a proportion of its force from escaping without the effect in the shape of smoke. To this end the efforts of scientific men will be more appropriately directed than to the vain design of inventing a substitute for coal, such as will enable us completely to dispense with that substance.

To Sunshine, then, we are indebted for coal, and throughout the economy of nature innumerable effects demonstrate its continuous influence.

Sunshine plays upon the surface of the land, the lakes, the rivers, and the oceans, and by the influence of its heat vapours ascend therefrom to form the clouds, which are to be condensed into rain in due season. The dews of night are born of sunshine, though fostered by cold. When the showers are withheld, then the sunshine hardens the mould, the earthy particles come into closer contact, and the mass becomes more solid, thereby preventing the heat and drought from penetrating the soil and killing both the seeds and the roots; for such would be the consequence if the soil did not become crusty and hard in dry weather.

If we be not warranted in concluding that the various

infallibly happen to Great Britain generally from the future exhaustion of its coal-fields. Emigration is the only alternative, sad as it may seem, and accordingly the process of comparative depopulation has begun in Cornwall. There may be still plenty of tin and copper in that county, but if in these days of free-trade they cannot be raised at a remunerative price, owing to foreign competition, then in a commercial point of view they may be regarded as worthless; and just so will it hereafter be with respect to our coal when a similar condition arrives.

We have adopted the above important considerations from a letter in the 'Times' of August 7, 1866, signed Y.

coloured rays which constitute white light owe their quality to the glowing vapour of the metals through which they pass, and which ultimately give the same property to the carbon of all artificial light,—still, it is certain that light is intimately connected with the phenomena of the colours that variegate creation. The colours of all bodies are either the simple colours of homogeneous light—violet, indigo, blue, green, yellow, and red (all reducible to red, blue, and yellow), or such compound colours as result from a mixture of homogeneous light. Whether colours result from reflected or transmitted light (as Delaval supposed), they are still produced by sunshine, by light; there is no colour in darkness. If solar light were of one colour only, all objects would appear of that colour only, or else black; but, as every ray of light is composed of all the colours of the rainbow, some things reflect one of these colours and some another. This effect results from the surface of things being differently constructed, both physically and chemically; and therefore some things reflect one ray, some two rays, some all the rays, and some none, but absorb all of them. Thus, the surface of the rose absorbs the blue and yellow rays of light, and reflects only the red. The surface of the violet absorbs the red and yellow rays of sunshine, and reflects the blue only. The primrose absorbs the blue and red rays of solar light and reflects the yellow : hence the respective colours of these flowers, and in like manner we may explain all the various tints of flowers. Those which are white reflect all the rays of light, absorbing none. Blackness, if it ever exists complete, results from the absorption of all the rays of light. White is thus the combination of all the colours ; black is the absence of all colour.

2 D

The green matter of all vegetable substances, termed
chlorophyll, is formed in their cells under the influence
of sunshine, and its green results from the fact that it
absorbs the red rays, reflecting both the blue and the
yellow together, which being mixed, as is well known,
produce green. In spring the green is light, because
the chlorophyll is not fully formed; in autumn it is a
yellowish-brown, when the chlorophyll decays. In the
two cases there is a deficiency of sunshine; but in the
full blaze of summer, as in the tropical regions, the
chlorophyll displays its lustrous and perfect green, so
pleasingly contrasting with the colour of the soil. In
fact, the chlorophyll can only be formed by sunshine:
hence, plants become a pale yellow when kept in the
dark. Bulbs, like the potato, which grow underground,
cannot form chlorophyll, because they are deprived of
sunshine: hence they are yellow, somewhat like plants
kept in the dark: but if potatoes are allowed to grow
out of ground, then sunshine developes the chlorophyll,
and they become green.

Green, in varying intensity, is the universal colour of
nature's garment, and our eyes and minds have accordingly been fashioned to find it most agreeable. We say
our eyes and minds have been fashioned to find this
universal green agreeable, for vegetation preceded the
advent of animals: so the eye and the mind which
judges its impressions, were adapted to the state of
things in which they were placed. The light—the sunshine which does not pass out from a body is in part
absorbed by that body, and *raises its temperature*; it
is also partly reflected, and thus renders the body and
its interior visible to the eye. Hence, the most visible of bodies are those which reflect most of the rays of

light,—the warmest of bodies are those which absorb most of the rays of light. Hence the coolness of white garments—as experienced in summer and in tropical regions—and the warmth of black or the darker colours in winter. The rise of temperature in all bodies from the absorption of light, according to their colour, is shown by placing woollen strips of different colours on snow, when the snow will be found to melt more or less under all the colours, but most under black, and not at all under pure white.

It is in the great vault of heaven that the effects of sunshine appear in all their grandeur. The atmosphere is one of the most transparent bodies known; when it is not loaded with fog, or obscured by other bodies, we can see objects placed at a very great distance; mountains do not disappear from our view until they are below the horizon. But, notwithstanding its feeble absorbing power, the air is not a body altogether transparent. If it were so, the vault of heaven would be black, and the sun and moon would appear as luminous disks accurately defined. At all places where the rays of the sun could not penetrate, either directly or reflected by terrestrial objects, there would be complete darkness; and, at the moment when the sun sets, night would suddenly succeed day. As all this does not happen, we must necessarily conclude that the particles of atmospheric air absorb a portion of the light they receive, allow a portion to pass, and reflect the third portion; hence it is that they illuminate the vault of heaven, light up terrestrial objects on which the sun does not shine directly, and determine an insensible and very gradual transition between day and night.

The blue colour of the atmosphere is the result of

sunshine. One part of the luminous rays is absorbed, the other reflected by the air; the air, however, does not act equally on all the coloured rays, of which light is composed; it acts like a milky glass—it rather allows the rays of the red extremity of the spectrum to pass, and, on the contrary, reflects the blue rays; but this difference is not sensible until the light has traversed *large masses* of air; hence, it is only in the large spaces of the celestial vault that air is blue—being in this respect like the ocean, the blueness of which results from the same cause, and only in masses of considerable depth.

The colour of the sky is modified by the combination of three tints: blue, which is reflected by the particles of air; the black of the vault of heaven, which forms the background of the atmosphere; and, finally, the white of the vesicles of fog and flakes of snow that swim in the atmosphere. Indeed, the tint of the blue rays is darkened by the black colour of space; and, on the other hand, it is made lighter by the white of the vesicles of fog. When we ascend in the atmosphere, we leave a great portion of the vesicles of fog-vapour beneath us; so that, while rays reach the eye in less proportion, the sky being covered with a lesser number of particles reflecting its light, its colour becomes of a deeper blue. For the same reason, the blue in the neighbourhood of the horizon is less intense than at the zenith. If the sky is paler on the open sea, and in high latitudes, than in the interior of the continents and in the neighbourhood of the equator, it must be attributed to the vesicles of fog.

Daybreak and twilight depend on the sunshine which meets the higher strata of the air, and, being reflected by it and dispersed in all directions, when the sun is ascending or has for some time disappeared below the horizon,—these rays are again reflected, and illuminate the western or eastern side, respectively, of the celestial vault.[1]

The duration and colouring of dawn and twilight depend on the state of the atmosphere. If the air be filled with vesicular vapour and the sky during the day has presented a whitish aspect, then the red is more or less dead and mingled with grey striæ or streaks, sometimes of a deep carmine colour; and even during the day, the part of the sky below the sun appears more or less red. It is therefore evident that these vapours are so arranged as to allow only the red rays to pass. Thus in winter, in our latitudes, the sky is frequently red throughout the day; and in summer, during rainy weather, when fine *cirrus* clouds are floating in the atmosphere, the same occurs several hours before the sun attains its greatest height; but when the sky has been of a deep blue throughout the day, then the twilight presents a yellowish tint. If light *cumulus* clouds or *cirrocumulus* are in the atmosphere, they are beautifully coloured, with green tints in the intervals between them.

The appearances of twilight depend on the sky and the rays of lingering sunshine; it therefore follows that they may serve to foretell, to a certain extent, the weather of the following day. When the sky is blue, and, after sunset, the western region is covered with a slight purple tint, we may be sure that the weather will be fine, especially if the horizon seems covered with a slight smoke. After rain, isolated clouds coloured red and and well-illuminated announce the return of fine weather. A twilight of a whitish-yellow, especially when

[1] See Chapter II. p. 12.

it extends to a distance on the sky, is not a sign of fine weather for the following day.

According to the opinion of country-people, we must expect storms when the sun is of a brilliant white, and sets in the midst of white light, which scarcely permits us to distinguish it. The prognostication is still worse when light *cirrus* clouds, which give the sky a dull appearance, appear deeper near the horizon, and when the twilight is of a *greenish-red*, in the midst of which are seen portions of a deep red, that pass into grey, and scarcely permit the sun to be distinguished : in this case, vesicular vapour is very abundant, and we may calculate on wind and approaching rain.

From daybreak the signs are somewhat deficient. When it is very red, as explained in a previous chapter, we may expect rain, whilst a grey morning announces fine weather. In addition to the general state of the vapour, as before explained, the reason of this difference between a grey dawn and a grey twilight is because, in the evening, this colour mainly depends on *cirrus* clouds —the icy clouds that condense the vapour, whilst in the morning it depends on the *stratus* cloud, which soon yields to the rising sun, whereas the *cirrus* cloud becomes thicker during the night. If at sunrise there is enough vapour for the sun to appear red, it is then very probable that, in the course of the day, the ascending current will determine the formation of a thick stratum of clouds, with consequent rain.

If we turn from the general appearances of nature, vegetation, the colours that adorn the earth and the skies, to the arts of life, and the conveniences of life, it is still sunshine to which we are indebted for all that is useful, agreeable, and necessary here below.

Throughout the entire range of chemistry, which ministers so largely to all our wants, the light and heat of the sun play incessantly a most important part. In some cases sunlight must be excluded, in others its action is indispensable, and both these conditions are wonderfully exemplified in the art of photography—the child of sunshine.

Photography depends upon the property of the salts of silver to be acted upon by light (or rather by the chemical or actinic rays accompanying rays of light) which darkens them according to its intensity, as reflected from the surface of bodies exposed to them.

But other metals, especially iron, are more or less endowed with the same sensibility to light; and probably all are so, only we have not as yet discovered the requisite solution to develop the impression, as in the case of silver and iron. With respect to the latter metal, the salts of it resulting from the combination with ammonia are highly sensitive, and may be utilized in photography. Papers imbued with ammonio-citrate of iron, for instance, will receive the photographic image, which may be developed by a solution of nitrate of silver into a presentable picture, more or less like that of the ordinary chloride-paper; whilst a solution of ferrocyanide of potassium produces from them pictures of a beautiful blue.

But the rays of sunshine are not only capable of chemical action; some of them are also magnetic,— endowed with the property of the loadstone,—capable of imparting magnetism to steel. Sometimes the figurative language of the poets hits off established facts in nature and science, long before the writers could have any idea of the real import of their words. Thus,

Milton plainly mentioned, although he did not know, the *magnetism of the solar rays* :—

> "The golden sun, in splendour likest heaven,
> Allured his eye . . . .
> . . . . . . where the great luminary
> (Aloof the vulgar constellations thick,
> That from his lordly eye keep distance due)
> Dispenses light from far. They, as they move
> Their starry dance, in numbers that compute
> Days, mouths, and years, towards his all-cheering lamp,
> Turn swift their various motions, or *are turn'd*
> *By his magnetic beam*, that gently warms
> The universe, and to each inward part,
> With gentle penetration, though unseen,
> *Shoots invisible virtue even to the deep*."[1]

What shall we say of this fine conception of our great poet now that the philosophers have ascertained, by direct experiment, that the *violet* ray of the solar spectrum is actually capable of rendering a needle magnetic which has never been touched by the loadstone or by an artificial magnet?   All that he meant to say, however, was that he supposed the earth revolved from west to east in consequence of a peculiar attraction (like that of a magnet) exerted on its substance by the sunbeams,— a beautiful poetical idea, and quite natural according to the state of science at his epoch.

In all the processes of dyeing, sunshine is an important agent.   In dyeing scarlet it is impossible to succeed perfectly without brilliant sunshine, as before stated.

If we expose an engraving to vivid sunshine for a few minutes, and then, taking it into a dark room, apply it to a piece of fresh photographic paper made sensitive by the chloride of silver, and keep the two in contact for a cer-

[1] 'Paradise Lost,' book iii.

tain time, we have only to apply the proper "developing solution" to see a *facsimile* of the engraving. Now, this proves beyond doubt that a chemical action has been set up by sunshine, and that this action has some duration after the cause has ceased. Light is, therefore, a material energetic thing, and its effect, in its totality, must be something prodigious throughout creation, the extent of which one cannot discover for want of the requisite "developing solution"! It is scarcely too much to believe that the images of all that exists are projected on all surfaces,—to pass away, as on the photographic film, because they are not "developed" and "fixed."

In the animal, as in the vegetable economy, sunshine is indispensable.

Vision is impaired by using the eyes with defective or deficient light; indeed, we seem warranted in concluding that light—sunlight—is absolutely necessary for the perfect development of the eyes. Fishes have been found in subterranean waters completely blind, but with "rudimentary" or undeveloped eyes,—indeed, without even a trace of eyes,—as in the underground waters of the Mammoth Cave, in America. Deficient light is a frequent cause of impaired eyesight at all ages, but no doubt the injury begins at the earliest ages, when the visual organs are being developed. Bloodless faces, sunken eyes, pupils dilated so as to diminish the extent of the iris or changing the colour to black, are the consequences of deficient light and sunshine; and the result is said to have occurred to those who have taken up their abode in the Mammoth Cave, before mentioned.

The narrow streets and lanes of cities must promote this evil to a great extent, by shutting out the light of the sky from our apartments. If, in a very narrow street or lane, we look out of a window with the eye in the same plane as the outer face of the wall in which the window is placed, we shall see the whole of the sky by which the apartment can be illuminated. If we now withdraw the eye inwards, we gradually lose sight of the sky till it wholly disappears, which may take place when the eye is only six or eight inches from its first position. In such a case, the apartment is illuminated only by the light reflected from the opposite wall, or the sides of the stones which form the window; because, if the glass of the window is six or eight inches within the wall, as it generally is, not a ray of light can fall upon it. Hence the deficient light and early darkness of such apartments.

The remedy is easy. If we remove our window, and substitute another in which all the panes of glass are *roughly ground on the outside,* and *flush with the outer wall,* the light from the whole of the visible sky, and from the remotest parts of the opposite wall, will be introduced into the apartment, reflected from the innumerable faces or facets which the rough grinding of the glass has produced. The whole window will appear as if the sky were beyond it, and from every point of this luminous surface light will radiate into all parts of the room.[1] We are persuaded that, by increasing the light by which we read, as we grow older, the necessity for spectacles might be considerably deferred.

If plants cannot grow without sunshine, animals pine away when deprived of it,—and none more than the young of human beings, to whom sunshine is absolutely necessary for healthy development. The elements of

[1] 'The Builder.'

our food which give rise to power and heat, are produced only by the action of sunlight.

Boys at their games are like young birds singing on the bough,—half of their joy is a result of the influence of the sunny air. The necessity of light for young children is not half appreciated. Many of the afflictions of children, and nearly all the cadaverous · looks of those brought up in great cities, are ascribable to this deficiency of light and air. When we see the glass rooms of the photographers in every street high up on the topmost story, we grudge them their application to a mere personal vanity. Why should not our nurseries be constructed in the same manner? If mothers knew the value of light to the skin in childhood, especially to children of a scrofulous tendency, we should have plenty of these glass-house nurseries, where children might run about in a proper temperature, free from much of that clothing which at present *seals up the skin*, —that great supplementary lung,—to sunlight and oxygen. Glass-house nurseries, lifted up to the topmost story, would save many a weakly child that now perishes for want of those necessaries of infant life.[1]

The rays of the sun become latent, so to speak, in animals and in plants, in the same way as the current of electricity becomes latent in the hydrogen by the decomposition of water. Man, by food, not only maintains the perfect structure of his body, but he daily lays in a store of power and heat, derived in the first instance from sunshine. The power and heat, latent for a time, reappear, and again become active when the living structures are resolved by the vital processes into their original elements. Finally, the rays of the sun add daily

[1] 'Pall Mall Gazette.'

to the store of indestructible forces in our terrestrial
body, maintaining life and motion. Every force in the
world—every power to lift a weight, from our own foot
off the ground to every ton lifted by a steam-engine, and
all the millions of tons of leaves and wood added to the
trees every year, and the clouds full of rain lifted from
the sea and rivers by evaporation, as much as the steam
off a boiling pot—is all due to heat; that is, ultimately
to the sun, to sunshine. In short, heat and force are
now well known to be convertible into each other.[1]
Thus, from beyond the limits of our earth, the body—
the mere earthy vessel—derives all that may be called
good in it; and of this not a single particle is ever lost.[2]
Such is what becomes of sunshine.

[1] Denison's 'Astronomy.'   [2] Liebig.

CHAPTER XXV.

WHAT BECOMES OF THE SHOWERS?

" WHAT's the use of all this rain?" is one of those questions the answer to which involves far more than we are apt to imagine. Such, certainly, is the case with the answer to the above question, and the investigation which it presupposes is one of the most interesting in the domain of science.

On the other hand, we may observe that the implication as to the *quantity* in " all this rain" is apt to be vastly exaggerated. The quantity of *water*, either in snow or rain, that annually falls on the surface of the earth is much less considerable than we fancy from the number of rainy days in the year, and even from those implacable downpours which sometimes last for entire days and nights in succession.

The " deluging" rains of the tropics—where it does, indeed, rain as nowhere else—do not positively produce more than eight inches of water in twelve hours; and in our latitudes, when we hear of floods and inundations, as recently from the heavy downpour in 1866, in various parts of the country, the absolute rainfall does not much exceed one inch and a quarter. The rain-gauge is the only test in this matter, and it dispels the delusion. But, indeed, according to M. Becquerel, as before stated, if the entire quantity of aqueous vapour in the atmosphere at any time were condensed into one diluvian downpour, it could not produce a depth of more than *four inches* of water on the surface of the globe. Surely this is more than enough to give us an idea of the small quantity of water that falls even in the heaviest showers. Moreover, we have ever and anon to complain of its deficiency, whilst surrounded by its great source, the world-encircling ocean.

But at Rajkote (Kattewas), India, during the month of July, 1849, it is stated that 26·75 inches of rain fell in 22 hours.

Mr. R. Strachan announced some curious facts relating to rainfall in 1862.[1] During the year the rainfall amounted to 25·67 in Gray's Inn Road, London. Rain fell on 179 days, that is, on nearly every other day. The *hours* of rain were estimated at 904; therefore, if the rain had fallen continuously, it would have lasted nearly 38 days and nights, although without the result of the 40 days which effected the Deluge, as discussed in a previous chapter (p. 100).

Still, the results of our rainfalls, when presented to us by calculation, assume striking proportions. Thus, a rainfall of 22 days in the month of October some years ago, in the counties of Kent, Sussex, and Hampshire, was calculated by Mr. Symons to represent in *weight* 2,729,636,160 tons—a *volume* equal to a solid mass whose base has an area of one square mile, and whose height is 3000 feet, say, one square mile and two-thirds of a mile high.

[1] 'Horological Journal,' vol. v.

A great portion of the surface of our planet is covered with water. The ocean drapes it round about as with a graceful garment, disclosing here and there much that is charming, and elsewhere what is anything but charming. If her surface be represented by 1, then the different seas united together would be expressed by 0·75, or, in plain language, three-quarters—without, however, including the various lakes, ponds, marshes, rivers, streams, and rivulets. Moreover, the dark interior of the earth herself contains an immense quantity of water, since we meet with water wherever we dig. In the cutting through marly clay for the railway from Tours to Bordeaux, between the valleys of the Cher and the Indre, M. Rozet discovered an immense quantity of *filets* of water, which crossed each other in every direction, and which might be compared to the veins in the body of an animal. Indeed, one cannot help considering water as the blood of the earth, which certain philosophers have considered a *living being*, with all the plants and animals preying upon her surface or skin, after the fashion of the parasites or very disagreeable insects which prey upon the skin of every animal. The soundings made for the construction of artesian wells, cr wells delivering water from great depths, show that there exist at very considerable depths immense masses and even currents of water; and it is these subterranean waters that give rise to all the watercourses on the surface, as well as the fountains and the lakes.

Now, as water is constantly changing into vapour at all temperatures—even when in the state of ice—all the water on the surface of the earth, and consequently in its interior, would become exhausted after a certain lapse of time, if there were no means of compensating its losses. This means is the precipitation of the vapour dispersed in the atmosphere—by dew, by snow, and by rain. As the level of the ocean, which is the great reservoir of the waters, has not varied for ages, as far as observation goes, it follows that the quantity of water continually carried off by evaporation is immediately restored by the precipitation of vapour. This fact involves the necessity of rain falling at all times upon a certain extent of the earth's surface; for the dew that falls every night could not alone compensate for evaporation. In effect, we know for certain that the seasons of sunshine in certain countries correspond to the seasons of showers in others, and *vice versâ*. Even in a country, rain is never or seldom general, any more than drought—at any rate, in our latitudes; hence, as previously observed, the absurdity of the almanac weather-predictions for every day in the year, without specifying the locality.

Rain, then, is indispensable for the maintenance of the universal equilibrium of the waters covering a portion of the surface of our planet, which nourish in their interior an immense number of living beings, plants, and animals, and serve to quench the thirst of those which live on the surface.

We all know the great influence of rain on the progress of vegetation, and that those countries which are deprived of it, without any watery compensation—such as the deserts of central Africa—are stricken with eternal sterility. It is not only by the moisture which it conveys to the soil that rain supports vegetation. Rain carries down with it a certain quantity of ammonia, from which the plants derive nitrogen—a gas which is indispensable to their growth. It also introduces into the soil the detritus of animals and plants, which decay, without any benefit to vegetation, in those countries where it never rains. By moistening the manure which the cultivator buries in the soil, rain facilitates its absorption by plants. Finally, it is probable that it is by means of the decomposition of the water which they absorb that plants procure a great portion of their hydrogen, if it be not indeed all their hydrogen. It is quite certain that more than 75 per cent. of all plants and animals consists of water; so that, after all, we may fairly be amazed at the thought of the size, the strength, the vigour, and all the other manifestations of plants, trees, and animals, man included, all which would vanish into nought were they deprived of the water in them, which is more than three-fourths of their entire substance!

Rain promotes the growth of the body. This seems at first sight incredible, but, since man includes in his composition the elements of the inferior natures, and among these the *vegetable,* it is probable that the very growth of our bodies may so depend upon moisture that it could not go on in air of a certain degree of dryness. It is at least evident, that men are of larger growth in *rainy* countries (whether these be warm or cold) than in those that are subject for a great part of the year to the dry extreme. In like manner, and from like causes in part, we see the inhabitants of crowded cities and manufacturing towns arrive at a less growth than those in even worse circumstances as to diet and clothing in the same country,—the latter being so much more exposed from childhood upwards to the weather.

The masses and watercourses underground, by reason of the capillarity and the property which water has of penetrating into all permeable bodies that touch it,

suffer continual losses from the quantity of water which
they send to the surface of the soil, on all the points
where there is no permeable layer, like potters' earth,
betwixt the surface and themselves.   Now, this is the
means which, in a great measure, supports vegetation
during drought.   Any one may convince himself of this
fact by delving under such parts of a field or piece of
ground in which the plants appear greener than in
others, when he will find the soil moister than elsewhere,
and moist even when the remainder is absolutely dry.
Evidently this moisture can be derived only from the
interior.   In all the oases of the deserts, where it never
rains, water is found by digging to a small depth; and
it is to its presence that those countries owe their beautiful vegetation in the midst of arid sands.   That water
certainly comes from the interior of the earth by currents issuing from the mountains which border the deserts; and these oases are points situated above those
currents, over which there happen to be no impermeable
layers between them and the soil, or above interior reservoirs in which water has accumulated, and which are
covered only with sand.

The aridity of deserts is not only due to the fact that
it never rains there, as has been supposed, but also to
the fact that an impermeable layer prevents the subterranean waters from mounting up to the surface of the soil.

It is generally believed that the subterranean waters
are supplied by those of the rains and snows which filter
gradually through the soil; but, according to M. Rozet's
experience and observation, it seems certain that rain-
water never penetrates to a great depth in our arable
soils which are permeable to water, but which have no
fissures.   It requires more than one day's continual rain

to moisten cultivated soil, after an ordinary drought, to the depth of eight inches; and after the heaviest rains, continued for several days in succession, the soil is not moistened beyond the depth of forty inches. But, indeed, this fact becomes at once apparent by observing the small depth of the layer of water that generally falls in a heavy shower. Moreover, all this water does not penetrate into the soil: a great portion of it runs on the surface to the rivulets, rivers, etc. A portion of that which penetrates into the earth is pumped up by the plants, which pour back again a great part of it into the atmosphere. The currents of air, in sweeping the surface of the soil, carry off from it, at every instant, a portion of the moisture which gets into it. Finally, the natural heat of the soil reduces into vapour a part of the water which it receives. From all these causes united, it happens that the water which penetrates through the particles of arable soils can never descend to a certain depth, and that a few dry days will suffice to exhaust it entirely.

It follows, therefore, that the water which supplies the subterranean reservoirs does not get there by penetrating the earth above them, as a mass of sand. It passes through the *fissures of rocks*, which are very numerous, especially in mountainous countries, being, as it were, nature's rain-gauges on the everlasting hills; hence the fact that such countries are situated betwixt the great reservoirs of the subterranean waters.

Certain meteorologists say that the mountains are regions of reservoirs because it rains more on mountains than on plains; but it happens to be just the contrary. It certainly rains *oftener* on mountains, but the annual rainfall of mountains is less than that of the plains at

2 E 2

their feet.   This is an incontestable fact, and it must be
evident even from the consideration of mere altitude or
height; all other conditions being equal, the greater the
height of a locality the less the rainfall as shown by the
rain-gauge.   The summits only get, as it were, a slice
of the showery column.

On the other hand, it appears that when rain takes
place with a turbid atmosphere, a considerable and variable proportion of the water is actually separated from
the vaporous medium at a height not exceeding 50 feet;
but in showers from an elevated region, falling through
an air which is not itself undergoing decomposition, the
products ought to be alike above and below.   Again,
during the heats of summer a portion of the rain falling
from an elevated region of the atmosphere is vaporized
near the surface by the heat of the latter.   If such be
the case, of course elevated gauges will give greater results than those on the ground.

It rains oftener on the mountains than on the plains
because their summits, in popular parlance, attract the
clouds.   Electricity, perhaps, plays a great part in this
phenomenon; but we are very far as yet from having
settled the point.   What is certain is, that the elevated
summits of mountains, radiating in all directions into
space, cool more than the soil of the plateaux situated
below them, whilst they are also frequently surrounded
with snows and glaciers.   They are, therefore, regions
of *minimum* temperature, which condenses the vapours
in their sphere of refrigeration, and compels them to
form *clouds*.   In Italy much less rain falls at the south
than at the north of the Apennines; and a vast deal
more falls at Bergen, at the foot of the Scandinavian
Alps, than at Chambéry, at the foot of Mont Blanc.

In generalizing on the discharge of districts in ordinary and flood time, the observer will be aware that there are seasons of drought when the most certain streams are seriously affected in their water-bearing properties. The water power of the Pentland Hills is very great, yet we have it on record that almost the whole of these sources were dried up in the summer of 1843. There appear to be, indeed, rare occasions, when hill-country suffers more from drought than lower districts; the lake district of Cumberland was similarly affected at about the same time.

In hotter countries these facts are proverbial. In the great tertiary plains of central Spain, which are 1200 feet and upwards above the sea, large rivers shrink into dry gravel-beds; while at the same time the seaward face of the mountains which fringe the Bay of Biscay are clothed with verdure caused by perpetual rains. This grand escarpment is broken up into the wildest and most precipitous glens, where the vapours rolling from the sea are caught and poured down with astonishing rapidity and enormous volume. The character of this great condensation, as it were, is so marked, that severe droughts are experienced on the inland side of the escarpment (5000 to 7000 feet above the sea) while within a very few miles rains are daily pouring down. These effects occur in a similar manner along the escarpment of the Bombay peninsula, where the rainfall is from 80 to 150 inches in the year, while it gradually recedes to 10 or 15 inches in the Deccan or dry country. The same is to be traced on the coast of Arracan, and in California on the western coast of America.[1]

[1] Beardmore, 'Hydraulic Tables.'

In the agricultural, or rather the national point of view, "what becomes of the showers" is a serious consideration; it is necessary to take account of the "appropriation" of rainfall.

The mean average quantity allowed for as disposed of by evaporation and absorption in the soil is very commonly taken at one-third of the whole depth,—often more in the south of England, sometimes exceeding a half;, whilst in the north generally one-third is ample, and often in excess of the actual amount.

The means of getting at these curious results is Dalton's Filtration and Evaporation Rain Gauge.* It consists of the usual funnel-mouth, which conveys the water into a cylinder some three inches in diameter, about three or four feet deep, and supported on an inverted funnel answering the purpose of a stand. The cylinder is filled with two or three feet deep of earth resting on a perforated bottom, through which the water reaching to that depth passes off into a short horizontal pipe, a little below a perforated plate, and connected with a vertical glass tube graduated, and bearing a definite proportion to the mouth of the gauge : by means of a tap the water may be drawn off when required. It is obvious that this arrangement is designed to imitate the general conditions of filtration or absorption, and the evaporation of rainfall. It was by a gauge similar to the one described that the following interesting results were obtained by Mr. Dickinson, at King's Langley, some years ago. The *entire* depth of rainfall was measured by the ordinary rain-gauge as given in the second column.

* The "Turf Gauge," mentioned in Symons's 'British Rainfall for 1865,' can scarcely be deemed an equivalent to Dalton's instrument.

Year.	Depth of Rainfall.	Filtration.	Evaporation.
1836	. . . 31·00	. . . 17·65	. . . 13·35
1837	. . . 21·10	. . . 6·95	. . . 14·15
1838	. . . 23·13	. . . 8·57	. . . 14·56
1839	. . . 31·28	. . . 14·91	. . . 16·39
1840	. . . 21·44	. . . 8·19	. . . 12·25
1841	. . . 32·10	. . . 14·90	. . . 17·91
1842	. . . 26·43	. . . 11·76	. . . 14·67
1843	. . . 26·47	. . . 8·10	. . . 15·37

Although such results, obtained in a single locality, cannot be accepted as a general rule, and may even, under certain conditions, be liable to exception,—still, experiments of this kind are very interesting, as tending to throw light to a certain extent upon a most important subject.

It is obvious that the nature of the land, and the climatic conditions of districts, must greatly influence the results of absorption and evaporation.

Districts may greatly vary in their general slope and geological character : grauwacke, granite, and the volcanic districts generally throw water in great rapidity, and are equally liable to great drought in summer time unless they are capped by moss-beds, which act as sponges not always the most pure ; some of the newer rocks, on the other hand, such as the old and new red sandstones, have great power of storing water;—the latter rocks, from their flatness, generally holding it, as indicated by the wells, which are always plentiful in this formation; the former, on the other hand, generally give out the purest spring-water when occurring on mountain slopes rising above the plains occupied by our numerous coal-fields. In the chalky districts this porous material absorbs a great portion of the rain it receives, collecting it in great underground sheets, represented by numerously interlaid flint-beds, and pouring out almost rivers at places that have no indication of a feeder; so strongly is this marked, that the chalk districts may be always identified upon the Ordnance maps by the absence of streamlets on its surface,—a characteristic likewise of some of the mountain limestones and oolites. This is the never-failing source of our Artesian wells.[1]

These instances are familiar to all who have studied the water-bearing properties of the hills of Great Britain : watching the progress of agriculture and drainage, we find the hill pastures scored in all directions with sheep-drains, while in agricultural districts thorough drainage steadily advances. All these operations are rapidly contributing to pour out floods on the dwellers in the plains, on the inhabitants of the rich levels at the mouth or on the lower course of our rivers, who, by this simple but irresistible order of events, find themselves forced into improvement which the natural resources of their soil had too long delayed. Frequently in this state of affairs the march of improvement has commenced in the harbour at the mouth, and the tidal volume is sent up vastly higher and sooner than known before; dredging-machines are set to work, and bold

---

[1] One of these, however, incomprehensibly fell one inch, "about three years ago." Unfortunately we could not ascertain the precise date of this fall. The Artesian well referred to is that of Messrs. Gordon and Co., Brewers, Caledonian Road, London,—one of the best Artesians existing. This magnificent well is 220 feet deep, where it reaches the chalk formation, after successively passing through yellow clay, blue clay, hard loamy sand, mottled clay, yellow sand, darker sand, dark loamy sand, dark grey sand and pebbles, and, finally, hard flints. At the greatest depths marine shells were discovered. The well yields 2,160 gallons of water per hour ; it is worked for fourteen hours daily ; and thus the daily yield is 30,240 gallons. At the Model Prison opposite, they went beyond the chalk in boring, and the consequence was they got no water.

piers or long river-walls are constructed, and, although
the landowner may find his outfall better, he also dis-
covers that a concurrent *high tide and upland flood* has
topped the walls and robbed his meadows of their
burthen, or swept down his ripening corn.[1]

The statistics of rain-fall, absorption, and evaporation
are not only valuable and interesting in a meteorological point of view, and for agricultural purposes, but
are highly important in connection with sanitary arrangements for towns and engineering operations. This
is especially evident to the hydraulic engineer. As
rain is an important source of water supply to rivers,
canals, reservoirs, it is evident that a knowledge of the
probable fall for any season or month at a given place,
as furnished by averages of the observations of former
years, will be the data upon which the engineer will
base his plans for providing for floods or droughts;
while the measurement of the actual quantity which has
just fallen, as gathered from the indications of a series
of gauges, will suggest to him precautions to adopt

---

[1] Beardmore, 'Hydraulic Tables.' The following formulæ may be
useful in estimating rainfall :—

1. *To ascertain the number of Cubic Feet of Water per Acre in
Rainfall.*—Multiply the constant, 363, by the depth of rain in *tenths* of
inches ; the product gives the cubic feet per square acre.   Example—
Rainfall, 4 in. ; then, 363 × 40 = 14,520 cubic feet per acre.

2. *To find the number of Cubic Feet per Square Mile.*—Multiply the
constant, 23232, by the rainfall in tenths of inches ; the product is the
answer.

3. *To find the Weight in Tons per Square Acre.*—Multiply the constant, 10128, by the rainfall in tenths of inches ; the product is the
weight nearly, with three places of decimals.

4. *To find the Weight in Tons per Square Mile.*—Multiply the constant, 64821, by the rainfall in tenths of inches ; the product is the
weight nearly, with two places of decimals.—Haskoll, 'Engineering
Field Work.'

either to economize or conduct away the in-pouring waters.[1]

The snow which at certain seasons covers the surface of the soil, finally melting below the limit of perpetual snow, also supplies the earth with a great quantity of water, which moistens it more than the rains. In fact, the melting of the snow being generally slow, the ground can continually absorb its products; whereas a great portion of the water from heavy showers flows over the soil, following the lines of the greatest slope towards the bed of rivulets and rivers, as before stated. On the other hand, the snow, remaining and preserving the surface which it covers from the contact of the air, prevents to a great extent the evaporation of the water which filters through; moreover, during the snow season the ground is much less covered with vegetation than during the rainy season, and we have just stated that plants continually carry off from the soil in which they live a notable portion of the water which it has absorbed.

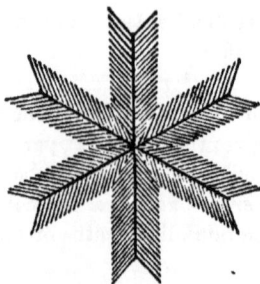

Snow Star.

Nothing can be more beautiful than the crystals of snow, to which we have before alluded. They vary

[1] Lardner, 'Natural Philosophy.'

from the simple union of six prisms in a *minute star*, as shown in the figure, to broad feathery flakes of the most compound structure. The figure is, of course, vastly enlarged in its likeness to the ordinary crystals, which are about the eighth of an inch in diameter. But during the late frost (1867) a correspondent to the 'Times' mentions the fall of snow-flakes which were of thin ice about the size of a small wafer, beautifully shaped like stars, each having *twelve* points diverging from the centre, in shape like the pricking part of a horseman's spur. Luke Howard also records the fall of large flakes of snow, among other instances, on the 8th of February, 1817, when the flakes were about an inch and a half in diameter.

It is not only by means of the water with which it furnishes it that snow is useful—very useful—to vegetation. Snow is also a huge screen interposed between the surface of the soil and the celestial spaces; it obstructs the radiation of the soil's heat into space; it enables the surface to cool much less rapidly than when uncovered by its woolly flakes; moreover, its white colour reduces the radiating power of the soil to the utmost. It is therefore certain that snow preserves the earth from cold during winter.

M. Rozet established this fact by direct and careful experiment during the last ten days of January, 1855, when the country round about Paris was covered with snow 2·34 inches in thickness. He found that the temperature of the air might diminish to a very great extent without the plants buried in the snow being in any danger from excessive cold. Moreover, during a thaw, the snow prevents them from passing suddenly from a low to a higher temperature, which is of itself the fre-

quent cause of destruction to many of them when unprotected by snow. Hence, in northern countries, and
in the regions of the Alps, the plants, especially the
cereals, emerge perfectly vigorous from beneath the
snow which has covered them for many months during
the hardest winters. It is well known that wheat and
many other plants vegetate under the snow at the very
time when similar plants are destroyed by the cold in
uncovered places. It is an established fact, in our latitudes, that those years whose winters are very snowy
yield the most abundant harvests; and that when trees
have been covered with hoar-frost during winter they
produce an abundant crop of fruit. It even appears,
from the experiments alluded to, that snow in the vicinity maintains the temperature of the soil which may be
uncovered, to a limited extent.

The surface of mountains cools more rapidly than
that of plains, and there are two causes of the difference:
first, their great elevation; and secondly, because, in
isolated places, they lose a great amount of heat by radiation. Now if, during winter, that surface remained
naked, the great intensity of the cold might destroy
both the plants which grow there naturally and those
cultivated by man. Thus, the snow preserves them from
this calamity.

Some such function as this—we mean the preservation of the earth's internal heat—must be subserved by
the *perpetual snows* of different regions. Those huge
isolated mountains would be vast radiators of the
earth's heat if the line of perpetual snow did not interpose its veto. So we find that this line gets lower
just in proportion to the necessity for preserving heat
in the different latitudes. Thus, it is highest on the

mountains near the equator (where there is plenty of
heat to spare), and then goes on descending lower
and lower, until, in fact, it has no elevation at all, snow
being perpetual on the surface in the Polar regions.[1]

On ordinary mountains the snow melts at the commencement of the warm seasons, and yields the water
which, trickling through the numerous fissures of the
rocks, goes to fill the subterranean reservoirs whence
issue the rivulets whose combination forms the rivers
which flow at the bottom of the valleys, making them
fruitful, and then flow in the plains, where man makes
use of them to water his plants, to give motion to
his mills, and to transport his produce, merchandise,
and bulky materials.

There is, without doubt, as we have felt it this year,
a grievous interruption of comfort in the invasion of
winter; but what if that is the best justification, the
greatest use, of its coming?   Fire and hail, snow and
vapour, wind and storm, have their errands, according to
the Psalmist.   The snow that cometh like wool, the
hoar-frost scattered like ashes, the ice cast forth like
morsels, and the frost which no one can abide, are but
servants in the great economy of the world.   Perhaps
there may be an advantage that, once a year, the great
god of this world, Comfort, whom we Englishmen—and
our American cousins perhaps to a still greater degree,
according to their lights—adore, should be dethroned.

---

[1] According to Humboldt, there is constant snow on mountains near
the equator at 15,700 feet; at 20° lat., 15,000 feet; at 45° lat., 8300
feet; and at 66° lat., 4900 feet.   According to M. Renou, the relation
of the perpetual snow-line is as follows :—*In all countries of the earth
the limit of perpetual snow is the altitude at which the hottest half of
the year has a mean temperature of* 32° *Fahr.*

We are brought face to face with facts we are too apt to
disregard; we are convinced of our insignificance in
the universe of beings; we may even be led, under the
pressure of common necessities, to a better sense of
common humanity.

Thus is the house of man built up by the Jack of
snow, in the order of Providence. Admirable is the
harmony that exists between all the phenomena which
we have described. As before observed, according to
the organization of living beings, all of which contain a
certain quantity of water, life could not be developed
without the intervention of this liquid. A vast layer of
water covers a portion of our globe, and masses more or
less considerable circulate in its interior. In order to
distribute this beneficent liquid to all organized beings
which are not plunged in it, the earth is surrounded by
a gaseous envelope—the atmosphere—designed to act as
a vehicle to water;.under the influence of the heat of
the sun and the proper heat of the earth, the water
rises slowly, and in the shape of molecules, into the
atmosphere, until the lowering of the temperature brings
it back into the solid state. Forces which are as yet
unknown to us, but among which are certainly electrical
influences, keep the vapour of water in the solid state,
and also in the vesicular state in suspension in the atmosphere, in the upper regions where the air is exceedingly rarefied. In these two states, vapour may form,
for a considerable time, homogeneous layers which are
not always visible below; but circumstances, still unknown to us, determine certain groupings in these layers
which finally produce *clouds*. It is a sort of summons
which bids the vapour to prepare to descend again into
reservoirs whence it went forth before.

Soon after the two species of clouds, the icy and the vesicular, are formed, they unite to descend together, at first in the shape of snow, which dissolves into rain on reaching the lower regions.*

A portion of the water which falls upon the earth goes directly into the ocean, the lakes, the watercourses; a second portion goes to the same destination after flowing over the soil; the third filters into the earth through the fissures of the rocks to go and compensate for the losses of the subterranean reservoirs; and, finally, the fourth is almost immediately restored to the atmosphere by means of the expiration of plants, and the evaporation caused by the proper heat of the earth.

Thus, there exists·a continual and universal circulation in the terrestrial reservoirs of water, even up towards the high regions of the air, and from there towards the same reservoirs. It is an endless chain in perpetual motion. This perpetual movement brings to each living being the quantity of water it needs, whilst, at the same time, it tempers the heat of the sun in summer and the intensity of cold in winter. Both, especially the former, would be intolerable without the moisture of the air.

" Water is, indeed, the best thing," exclaimed Pindar of old, little dreaming how completely science would explain and corroborate the axiom. If it may be doubted, as it is by some, that no living thing exists

---

* According to the observations of M. Silbermann, recently put forth, storm-clouds always originate in an aggregation of a great number of other clouds in the form of cumulo-stratus. From this union of masses, at first isolated, there results one large cloud in the form of a mushroom, somewhat like a tree resting on a wide base of cumulo-stratus. It is always in the middle of the part immediately above the trunk, in the bosom of the foliage of this sort of electric tree, that the focus of the lightning seems to exist.

without having been produced by an egg, or its equivalent, it.is certain that water is indispensable to vitality of every kind—to all living organisms. The origin of all animals as well as plants is traced to "the watery element." In the simple narrative of Genesis, the all-important function of "the waters" is abundantly evident.

Present at the evolution of every germ, water nurtures growth or brings about the endless changes which constitute the real metamorphosis of creation.

Finally, in the showers, water is the universal scavenger of nature. The showers *purify the air*. They beat down the noxious exhalations collected in it, and dissolve them. They mix the air of the upper regions with that of the lower. They *wash the earth*, even as mothers wash their infants. They set in motion the stagnant contents of ditches and sewers, and mix them with purifying earth, or waft them to the universal purifier and disinfectant—the ever defiled, but still constantly clean, Ocean.

Such is the simplicity of the mechanism which gives and secures existence to so many millions of organized beings ; and we have, literally, only to open our eyes to see it in operation, and to comprehend it in its totality. The least we can say is—Thank God for it !

THE END.

PRINTED BY J. E. TAYLOR AND CO.,
LITTLE QUEEN STREET, LINCOLN'S INN FIELDS.